AF581856

Effects of Diagenetic Alterations on Hydrocarbon Reservoirs and Water Aquifers

Effects of Diagenetic Alterations on Hydrocarbon Reservoirs and Water Aquifers

Editor

Howri Mansurbeg

MDPI • Basel • Beijing • Wuhan • Barcelona • Belgrade • Manchester • Tokyo • Cluj • Tianjin

Editor
Howri Mansurbeg
Salahaddin University-Erbil
Iraq
Palacký University Olomouc
Czech Republic

Editorial Office
MDPI
St. Alban-Anlage 66
4052 Basel, Switzerland

This is a reprint of articles from the Special Issue published online in the open access journal *Water* (ISSN 2073-4441) (available at: https://www.mdpi.com/journal/water/special_issues/diagenetic_reservoirs_water_aquifers).

For citation purposes, cite each article independently as indicated on the article page online and as indicated below:

LastName, A.A.; LastName, B.B.; LastName, C.C. Article Title. *Journal Name* **Year**, *Volume Number*, Page Range.

ISBN 978-3-0365-5545-4 (Hbk)
ISBN 978-3-0365-5546-1 (PDF)

Cover image courtesy of Howri Mansurbeg

Contents

About the Editor

Howri Mansurbeg

Howri Mansurbeg is a Sedimentary Geologist with expertise in diagenesis, sequence stratigraphy, and reservoir characterization of siliciclastics and carbonates. He has obtained his PhD from Uppsala University, Sweden, in 2007. His current research deals with hydrocarbon exploration and exploitation in carbonate reservoirs in Iraqi Kurdistan and the United Arab Emirates, utilizing an integrated multidisciplinary approach. Howri has worked for several international oil companies in Europe and the US and has produced many technical reports, research articles, and book chapters in the past 15 years. He is currently a Professor at the General Directorate of Scientific Research Center, Salahaddin University-Erbil, the Kurdistan region of Iraq and a Senior Researcher at the Department of Geology, Faculty of Science, Palacký University Olomouc, Czech Republic.

Editorial

Effects of Diagenetic Alterations on Hydrocarbon Reservoirs and Water Aquifers

Howri Mansurbeg [1,2]

[1] General Directorate of Scientific Research Center, Salahaddin University-Erbil, Erbil 44002, The Kurdistan Region of Iraq, Iraq
[2] Department of Geology, Faculty of Science, Palacký University Olomouc, 17. listopadu 1192/12, 771 46 Olomouc, Czech Republic

Citation: Mansurbeg, H. Effects of Diagenetic Alterations on Hydrocarbon Reservoirs and Water Aquifers. *Water* **2022**, *14*, 2915. https://doi.org/10.3390/w14182915

Received: 6 September 2022
Accepted: 13 September 2022
Published: 17 September 2022

Diagenesis includes all the biological, physical, chemical, biochemical, and physicochemical alterations that occur immediately after deposition and prior to low-grade metamorphism. These processes exert direct control on the hydrocarbon reservoir quality and on the hydraulic properties of groundwater aquifers [1]. As stated by Ahr (2008), "Reservoirs and aquifers differ only in the fluids they contain" [2]. Thus, methods to assess the geological storage and transmission capacity of hydrocarbon reservoirs are applicable to groundwater aquifers [3]. Overall, diagenetic alterations are, directly or indirectly, driven and mediated by fluid flows in sedimentary basins that determine the porosity and permeability evolution of carbonate and sandstone reservoirs/aquifers. The pervasiveness and extent of cementation in any reservoir/aquifer are determined by the rate and volume of fluids percolating through these sediments, temperature and geochemical conditions, and the time available for cementation [4]. Thus, publishing papers dealing with diagenesis and reservoir quality evolution in MDPI's *Water* is highly relevant and a welcome approach.

Moreover, reservoir/aquifer layers (carbonates and sandstones) experience several phases of diagenesis, which are related to the tectonic history of the basin. Episodes of subsidence and uplift may induce dramatic changes in the pressure–temperature realms and in formation water chemistry that are reflected in various phases of mechanical and chemical compactions as well as mineral dissolution, transformation, and cementation [1,5]. Diagenesis may result in deterioration through mechanical and chemical compaction, preservation through the prevention of mechanical compaction and cementation, or enhancement due to the dissolution of allochems (framework grains) and intergranular cements [2,6–9].

This Special Issue, "Effects of Diagenetic Alterations on Hydrocarbon Reservoirs and Water Aquifers", is a collection of five papers that document new geological, geochemical, and sedimentological data from basins in the Middle East, Canada, and China.

The paper by Salih et al. [10], entitled "Hydrothermal Fluids and Cold Meteoric Waters along Tectonic-Controlled Open Spaces in Upper Cretaceous Carbonate Rocks, NE-Iraq: Scanning Data from In Situ U-Pb Geochronology and Microthermometry", provides petrographic and geochemical data on Upper Cretaceous carbonate rocks in NE Iraq. These authors documented the role of high-temperature and hydrothermal brines in the precipitation of saddle dolomite cement within the fractures. The in situ U-Pb dating method was used to date the age of the saddle dolomites, and it linked their dating to the tectonic evolution of the Zagros Foreland Basin.

The second paper by Salih et al. [11], entitled "Tracking the Origin and Evolution of Diagenetic Fluids of Upper Jurassic Carbonate Rocks in the Zagros Thrust Fold Belt, NE-Iraq", studied the evolution of diagenetic fluids from the Upper Jurassic Barsarin Formation in the Zagros Foreland Basin in Iraqi Kurdistan. The study was based on fieldwork, petrography, carbon and oxygen isotopes, and in situ strontium isotope ratios using high-resolution laser ablation ICP-MS. The study provides a detailed account of the evolution of different fluids during the diagenesis of the Upper Jurassic Barsarin Formation and how the fluids shifted from the original chemistry and affected the host carbonate rocks. The study also provides insights into the dynamics of the diagenetic fluids during the Zagros orogeny within the formation of the Zagros Foreland Basin.

The paper by Al-Aasm et al. [12], entitled "Dolomitization of Paleozoic Successions, Huron Domain of Southern Ontario, Canada: Fluid Flow and Dolomite Evolution", integrated petrographic, isotopic, fluid inclusion microthermometry, and geochemical analyses of Paleozoic carbonate successions from multiple boreholes within the Huron Domain, Southern Ontario, in order to shed light on the nature of the diagenetic alterations and the chemical composition of the fluids responsible for causing dolomitization on a regional scale. This study also evaluated the nature and origin of dolomitized beds in the successions studied. Detailed geochemical analyses of, for example, major (Ca, Mg), minor, trace (Sr, Na, Mn, and Fe), and rare-earth elements (REEs), along with the use of an environmental scanning electron microscope (ESEM) coupled with an EDAX detector, were utilized to analyze selected carbon-coated thin-section samples to investigate the nature of the dolomitization and subsequent recrystallization of the precursor dolomite matrix.

The paper by Mansurbeg et al. [13] addresses the controversies regarding the nature and origin of drusy mosaic dolomite cement, which commonly occludes pores in carbonate reservoirs. The study sheds light on the nature and origin of the drusy dolomite in terms of whether it has been directly precipitated as cement or formed by the replacement of precursor drusy calcite. The study unravels the role of diagenetic paleofluids in forming drusy mosaic dolomite cement and has implications for accurate construction of paragenetic sequences and related reservoir quality features.

The most important diagenetic alteration in quartz-rich sandstones is the precipitation of syntaxial quartz overgrowths around detrital quartz grains. Deeply buried sandstone reservoirs lose primary porosity through the process of quartz cementation. In the voluminous literature on quartz cementation in sandstone reservoirs, the source of silica needed for quartz cementation is uncertain, and the issue is strongly debated among researchers. The paper by Ren et al. [14] used high-precision secondary ion mass spectrometry (SIMS) for micrometer-sized quartz cement in order to obtain oxygen isotopes. The obtained $\delta^{18}O$ values coupled with petrographic, cathodoluminescence (CL), and fluid inclusion data were used to constrain the origin of the silica in quartz cement in the deep-buried Upper Triassic tight sandstones in the western Sichuan Basin, China.

There is mounting evidence of the presence of unextracted conventional and unconventional hydrocarbon reserves and untapped groundwater aquifers in deeply buried and complex diagenetic regimes and tectonic settings. The current predictive tools and concepts to predict reservoir/aquifer quality in petroleum and groundwater exploration have limitations. Further studies are necessary for the development of multidisciplinary diagenetic models at the basin scale coupled with geological processes operating on the reservoirs after deposition. These studies might contribute to a better understanding of the parameters controlling reservoir/aquifer quality evolution and help in developing more accurate predictive tools.

Funding: This research received no external funding.

Acknowledgments: I thank the authors of the articles included in this Special Issue and the organizations that have financially supported the research in the areas related to this topic. I thank Section Managing Editor Shanika Zhu for the tremendous support and guidance.

Conflicts of Interest: The author declares no conflict of interest.

References

1. Morad, S.; Al-Ramadan, K.; Ketzer, J.M.; De Ros, L.F. The impact of diagenesis on the heterogeneity of sandstone reservoirs: A review of the role of depositional facies and sequence stratigraphy. *AAPG Bull.* **2010**, *94*, 1267–1309. [CrossRef]
2. Ahr, W.M. *Geology of Carbonate Reservoirs: The Identification, Description, and Characterization of Hydrocarbon Reservoirs in Carbonate Rocks*; Wiley: Hoboken, NJ, USA, 2008; pp. 1–277.

3. Bloomfield, J.P. The role of diagenesis in the hydrogeological stratification of carbonate aquifers: An example from the Chalk at Fair Cross, Berkshire, UK. *Hydrol. Earth Syst. Sci.* **1997**, *1*, 19–33. [CrossRef]
4. Harris, P.M. Depositional environments of carbonate platforms. *Colo. Sch. Mines Q.* **1986**, *80*, 31–60.
5. Swart, P.K. The Geochemistry of Carbonate Diagenesis: The Past, Present and Future. *Sedimentology* **2015**, *62*, 1233–1304. [CrossRef]
6. Kupecz, J.A.; Gluyas, J.; Bloch, S. (Eds.) Reservoir quality prediction in sandstones and carbonates: An overview. *Am. Assoc. Pet. Geol. Mem.* **1997**, *69*, 311.
7. Moore, C.H.; Wade, W.J. *Carbonate Reservoirs. Porosity and Diagenesis in a Sequence Stratigraphic Framework*, 2nd ed.; Elsevier: Amsterdam, The Netherlands, 2013.
8. Esch, W.L. Multimineral diagenetic forward modeling for reservoir quality prediction in complex siliciclastic reservoirs. *Am. Assoc. Pet. Geol.* **2019**, *103*, 2807–2834. [CrossRef]
9. Ehrenberg, S.N.; Zhang, J.; Gomes, J.S. Regional porosity variation in Thamama-B reservoirs of Abu Dhabi. *Mar. Pet. Geol.* **2020**, *114*, 104245. [CrossRef]
10. Salih, N.; Mansurbeg, H.; Muchez, P.; Gerdes, A.; Préat, A. Hydrothermal Fluids and Cold Meteoric Waters along Tectonic-Controlled Open Spaces in Upper Cretaceous Carbonate Rocks, NE-Iraq: Scanning Data from In Situ U-Pb Geochronology and Microthermometry. *Water* **2021**, *13*, 3559. [CrossRef]
11. Salih, N.; Préat, A.; Gerdes, A.; Konhauser, K.; Proust, J.-N. Tracking the Origin and Evolution of Diagenetic Fluids of Upper Jurassic Carbonate Rocks in the Zagros Thrust Fold Belt, NE-Iraq. *Water* **2021**, *13*, 3284. [CrossRef]
12. Al-Aasm, I.S.; Crowe, R.; Tortola, M. Dolomitization of Paleozoic Successions, Huron Domain of Southern Ontario, Canada: Fluid Flow and Dolomite Evolution. *Water* **2021**, *13*, 2449. [CrossRef]
13. Mansurbeg, H.; Alsuwaidi, M.; Dong, S.; Shahrokhi, S.; Morad, S. Origin of Drusy Dolomite Cement in Permo-Triassic Dolostones, Northern United Arab Emirates. *Water* **2021**, *13*, 1908. [CrossRef]
14. Ren, J.; Lv, Z.; Wang, H.; Wu, J.; Zhang, S. The Origin of Quartz Cement in the Upper Triassic Second Member of the Xujiahe Formation Sandstones, Western Sichuan Basin, China. *Water* **2021**, *13*, 1890. [CrossRef]

Article

Hydrothermal Fluids and Cold Meteoric Waters along Tectonic-Controlled Open Spaces in Upper Cretaceous Carbonate Rocks, NE-Iraq: Scanning Data from In Situ U-Pb Geochronology and Microthermometry

Namam Salih [1,2,*], Howri Mansurbeg [3,4], Philippe Muchez [5], Axel Gerdes [6,7] and Alain Préat [8]

1 Petroleum Engineering Department, Engineering Faculty, Soran University, Soran-Erbil 44008, The Kurdistan Region, Iraq
2 Scientific Research Centre (SRC), Soran University, Soran-Erbil 44008, The Kurdistan Region, Iraq
3 General Directorate of Scientific Research Center, Salahaddin University-Erbil, Erbil 44001, The Kurdistan Region, Iraq; howri.mansurbeg@gmail.com
4 School of the Environment, University of Windsor, Windsor, ON N9B 3P4, Canada
5 KU Leuven, Department of Earth and Environmental Sciences, Celestijnenlaan 200E, B-3001 Leuven, Belgium; philippe.muchez@kuleuven.be
6 Institute for Geosciences, Goethe-University Frankfurt am Main, Altenhöferallee 1, 60438 Frankfurt, Germany; gerdes@em.uni-frankfurt.de
7 Frankfurt Isotope and Element Research Center (FIERCE), Goethe-University Frankfurt, 60323 Frankfurt, Germany
8 Research Group.—Biogeochemistry & Modelling of the Earth System, Université Libre de Bruxelles, 1050 Brussels, Belgium; alain.preat@ulb.be
* Correspondence: namam.salih@soran.edu.iq; Tel.: +964-750-460-5345

Citation: Salih, N.; Mansurbeg, H.; Muchez, P.; Gerdes, A.; Préat, A. Hydrothermal Fluids and Cold Meteoric Waters along Tectonic-Controlled Open Spaces in Upper Cretaceous Carbonate Rocks, NE-Iraq: Scanning Data from In Situ U-Pb Geochronology and Microthermometry. *Water* **2021**, *13*, 3559. https://doi.org/10.3390/w13243559

Academic Editor: Thomas Meixner

Received: 18 October 2021
Accepted: 6 December 2021
Published: 13 December 2021

Abstract: The Upper Cretaceous carbonates along the Zagros thrust-fold belt "Harir-Safin anticlines" experienced extensive hot brine fluids that produced several phases of hydrothermal cements, including saddle dolomites. Detailed fluid inclusion microthermometry data show that saddle dolomites precipitated from hydrothermal (83–160 °C) and saline fluids (up to 25 eq. wt.% NaCl; i.e., seven times higher than the seawater salinity). The fluids interacted with brine/rocks during their circulation before invading the Upper Cretaceous carbonates. Two entrapment episodes (early and late) of FIs from the hydrothermal "HT" cements are recognized. The early episode is linked to fault-related fractures and was contemporaneous with the precipitation of the HT cements. The fluid inclusions leaked and were refilled during a later diagenetic phase. The late episode is consistent with low saline fluids (0.18 and 2.57 eq. wt.% NaCl) which had a meteoric origin. Utilizing the laser ablation U-Pb age dating method, two numerical absolute ages of ~70 Ma and 3.8 Ma are identified from calcrete levels in the Upper Cretaceous carbonates. These two ages obtained in the same level of calcrete indicate that this unit was twice exposed to subaerial conditions. The earlier exposure was associated with alveolar and other diagenetic features, such as dissolution, micritization, cementation, while the second calcrete level is associated with laminae, pisolitic, and microstromatolite features which formed during the regional uplifting of the area in Pliocene times. In conclusion, the hydrothermal-saddle dolomites were precipitated from high temperature saline fluids, while calcrete levels entrapped large monophase with very low salinity fluid inclusions, indicative for a low temperature precipitation from water with a meteoric origin.

Keywords: hydrothermal fluids; cold meteoric waters; Upper Cretaceous carbonate rocks; U-Pb geochronology; fluid inclusion microthermometry

1. Introduction

Hydrothermal fluids and meteoric waters are characterized by a significant heterogeneity in geochemical composition. The heterogeneities in composition bring a considerable challenge for utilizing more sophisticated tools to decipher the origin of the fluids involved during diagenesis, and the burial and temperature history of hydrocarbon reservoirs. To gain better insights into the origin and evolution of these diagenetic fluids, techniques such as conventional petrography, cathodoluminescence, and stable and radiogenic isotopes have been applied [1–6]. The researchers used these methods to understand the complex diagenetic settings in hydrothermal dolomitization. [7] followed a similar approach to identify the conditions which lead to the calcretization of an Upper Cretaceous series in relation with the timing of a dolomitization sequence which was affected by various hot and cold fluids that caused numerous heterogeneities (sensu [8]). Interconnected factors caused a complex heterogeneity in hydrothermal fluids "HT" and cold meteoric waters, including: the source of the fluid's influx/reflux into the host rocks; the amount of hot or cold waters; tectonics; depositional settings; and the interaction of these fluids, especially HT fluids with other stratigraphic sequences prior to the circulation through the final host rocks—all of these processes led to generate a complex pattern [9,10].

This study deals with the origin and composition of the fluids flowing along deep-seated faults through the Upper Cretaceous host rocks of the Bekhme formation (Figure 1). In HT fluids, usually the researchers follow the classical studies, starting from a wide descriptive scenario with analytical techniques to draw the conceptual models for understanding the relative timing of these fluids/waters or their origin [11]. However, the interpretation of these HT fluids remains controversial and fundamental questions remain unanswered. Previously and some recent studies interpreted HT as one phase based on general stratigraphy and structures of the studied area [8,12], while a recent study estimated the absolute age of these fluids by utilizing a laser ablation technique (U-Pb dating [7]). The crucial question, which needs elaboration is the nature and distribution of several HT phases that took place through the circulation of HT fluids with significant variation of salinities. One of the major questions is to determine the number of pulses and the injection age of these HT fluids, in order to unravel the possible driving mechanisms for their upward flow [13].

This study in the Upper Cretaceous Bekhme carbonate formation along Harir-Spelek Basin in the Zagros fold-thrust belt (ZFTB) will add new data to the previous studies in the same area [7,8] to follow the origin of these HT fluids by integrating conventional petrography (including cathodoluminescence) with detailed fluid inclusion petrography, microthermometry, laser ablation "LA" U-Pb dating, and stable isotopes, to constraint paleotemperatures and paleosalinities of the dolomitizing fluids. U-Pb dating is particularly useful to determine the pulse chronology and their possible sources.

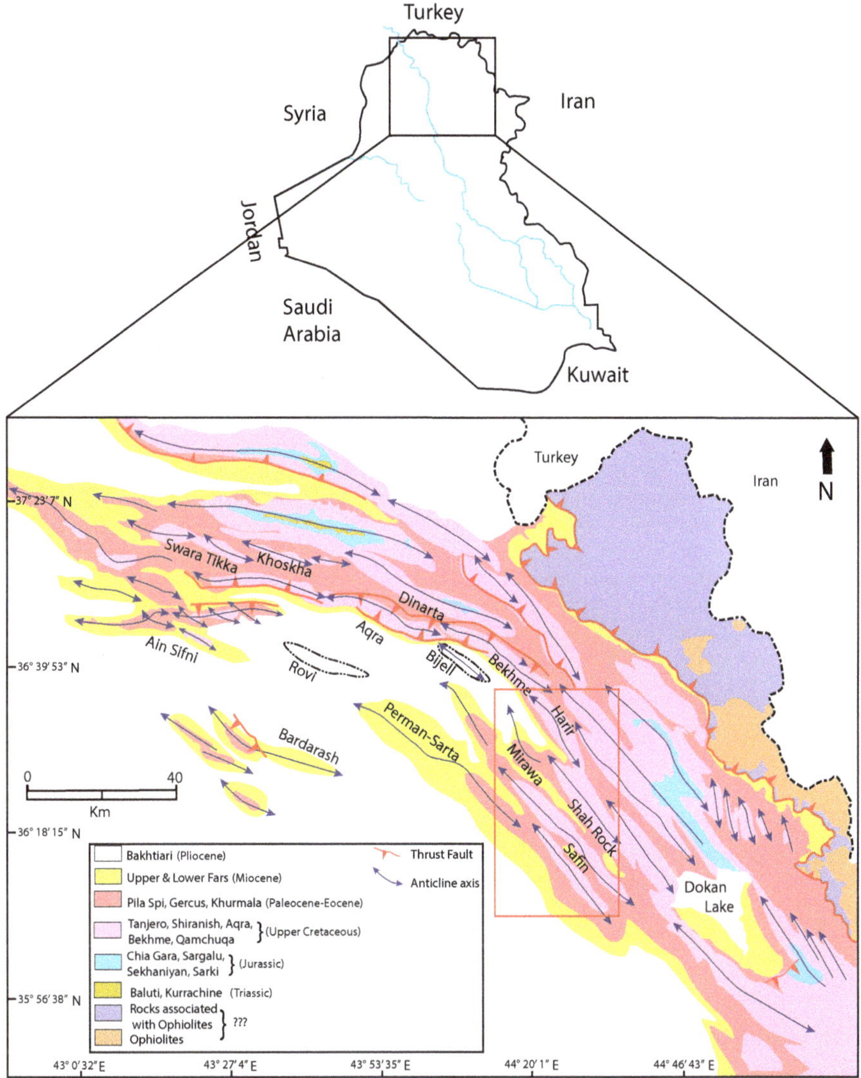

Figure 1. Geological and tectonic map showing the general location of the studied areas "Spelek-Sulauk", see red rectangle (after [7]).

2. Geological Setting

The study area is located within the Zagros fold-thrust belt (ZFTB), Kurdistan Region, northeast Iraq. The area is considered as an important anticlinal structure in the high folded zone, called Harir and Safin anticlines (Figure 1). The study area consists mostly of Upper Cretaceous carbonates at the core of Harir and Safin anticlines, with shallow-water environmental settings [14]. One of the most obvious formations within Cretaceous rocks is the Bekhme formation, which was deposited in a marine shelf environment with rudists, and benthic and planktonic foraminifera [15]. The Bekhme formation is one of the carbonate oil reservoirs in the subsurface and is well exposed to the surface.

During Permian, the opening of the Neotethys oceans started, and was associated with rifting along the ZFTB. The latter underwent complex tectonic events with extensional phases along both passive margins in the Late Triassic [16]. These extensional phases were followed by the Jurassic–Miocene arc-continent collision and ophiolite obduction along the Arabian plate [17]. Finally, post-orogenic strike-slip faulting progressed and later on the segmentation of Harir and Safin anticlines occurred. The area can be divided into four blocks with a general NW–SE trend. The blocks were controlled tectonically by basement movement along a detachment surface. The strike-slip/oblique faults with a NE–SW trending direction of the Harir and Safin anticlines could reflect the development of a post-orogenic faulting system after the Arabian and Eurasian plate collision.

During the Upper Cretaceous when Bekhme formation was deposited, the basin went through intense tectonic activities, which resulted from the collision between the Arabian and Eurasian plates. The overall tectonic setting shifted from a passive margin to a foredeep setting [18]. The shift in the tectonic regime had a significant impact on the depositional settings and the nature of the sedimentary rocks deposited [18].

3. Materials and Methods

Fluid inclusion petrography and microthermometry were performed on double polished wafers (100–120 μm thick). The wafers were studied under transmitted and fluorescent light in order to distinguish the type of fluid inclusions. Measurements were performed on a Linkam MDS-600 heating-freezing stage(Linkam Scientific Instruments Ltd., Waterfiels, Epsom, Tadworth, UK) mounted on an Olympus BX51 microscope (Olympus Corporation, Tokyo, Japan). The stage was calibrated by measuring synthetic Syn FlincTM fluid inclusion standards (Fluid Inc, Denver, CO, USA). The occurrence of about 110 primary and secondary fluid inclusions were studied under the microscope. The temperatures were obtained by heating and cooling the inclusions in a chamber mounted on a microscope (Table 1). The fluid inclusions present contained two phases, i.e., a liquid (L) and a vapor phase (V). A minimum temperature of the fluid from which a specific mineral precipitated could be obtained during heating. The temperature at which the liquid and vapor phases homogenize is called the homogenization temperature (Th), and is the minimum temperature of entrapment of the fluid inclusion. Freezing and subsequent heating fluid inclusions allow to determine the temperature of first melting of aqueous solid phase and of the final melting. Based on these melting temperatures, the major composition and the salinity of the ambient fluid can be deduced, respectively. The heating rate during homogenization of the fluid inclusions was 3 °C/min and 1 °C/min at the phase transitions during melting. The study and analyses were performed at KU Leuven (Belgium). The preparation technique and measurement procedure have been described by Muchez [19].

Table 1. Distribution and microthermometric data from fluid inclusions. Sp. = Spelek section. Sua. and Sub. = Sulauk sections. Th = homogenization temperature. Tfm = first melting temperature. Tf = freezing temperature. Tmice = final melting temperature of ice. NE-Iraq-Kurdistan Region.

Location	Sample	Rock Phase	Th (°C)	Tf (°C)	Tfm (°C)	Tmice (°C)	Salinity%
Spelek	Sp.16	Saddle Dolomite (SD2)	96	−82	-	−22.3	24.7
			100	−76	-	−20.5	22.7
			100	−76	-	−21.8	23.5
			102	−58	-	−11.9	15.9
			150	−67	-	−23.5	24.7
			150	−65	-	−22.0	23.7
			113	−66	-	−22.4	24

Table 1. *Cont.*

Location	Sample	Rock Phase	Th (°C)	Tf (°C)	Tfm (°C)	Tmice (°C)	Salinity%
			150	−65	-	−22.0	23.7
			113	−66	-	−22.4	24
			102	−69	-	−22.2	23.9
			110	−74	-	−23.0	24.3
			116	−66	-	−21.1	23.8
			86	−74	-	-	-
			120	−69	-	-	-
			121	-	-	-	-
			125	-	-	-	-
			99	-	-	-	-
			141	-	-	-	-
	Sp.19	Saddle Dolomite (SD2)	96	−71	-	−19.7	22.2
			92	−75	−48.6	−20.2	22.5
			94	−74	−45.7	−20.3	22.6
			140	−69	-	−20.3	22.6
			94	-	-	-	-
			94	-	-	-	-
			103	-	-	-	-
			110	-	-	-	-
			>150	-	-	-	-
			94	-	-	-	-
			116	-	-	-	-
			99	-	-	-	-
			99	-	-	-	-
			102	-	-	-	-
			95	−75	-	−19.3	21.7
			86	-	-	-	-
			99	−68	-	-	-
			100	−68	-	-	-
Sulauk	Sub.2	Late saddle dolomite (SD3)	109	−60	-	−14.0	17.8
			>122		-	-	-
			110		-	-	-
			98		-	-	-
			118	−82	-	−19.8	22.2
			116	−78	-	−18.8	21.5
			118	−82	-	−18.8	21.5
			118	−86	-	−21.5	23.4
			119	−83	-	−21.8	23.5
			99		-	-	-
			94	-	-	-	-

Table 1. *Cont.*

Location	Sample	Rock Phase	Th (°C)	Tf (°C)	Tfm (°C)	Tmice (°C)	Salinity%
			96	-	-	-	-
			122	-	-	-	-
			98	-	-	-	-
			98	-	-	-	-
			99	-	-	-	-
			95	-	-	-	-
			109	-	-	-	-
			99	-	-	-	-
			104	-	-	-	-
			122	-	-	-	-
			96	-	-	-	-
	Sua.3 and Sua.5	Saddle dolomite (SD3)	107	−71	-	−20.2	22.5
			100	−74	-	−20.3	22.6
			106	−72	-	−20.5	22.7
			117	-	-	-	-
			108	−62	-	−11.8	15.8
			108	−62	-	−12.2	16.2
			118	−62	−39.8	−12.2	16.2
			109	−62	-	−11.9	15.9
			96	-	-	-	-
			110	-	-	-	-
			86	−63	-	−15.2	18.8
			97	-	-	-	-
			109	-	-	-	-
			110	−60	-	−17.8	20.8
			113	-	-	-	-
			104	−70	-	−20.5	22.7
			102	−73	-	−20.5	22.7
			100	−62	-	-	-
			100	−62	-	-	-
			110	−62	-	−13.7	17.5
			110	−62	-	−13.8	17.6
			110	−62	-	−13.8	17.6
			128	−62	-	−16.8	20.1
			86	−82	−41.1	−21.5	23.4
			91	−79	−41.8	−21.6	23.4
			91	−81	-	−21.9	23.6
			86	−78	−41.4	−16.6	19.9
			88	−81	-	−20.8	22.9
			92	−75	-	−17.9	20.9

Table 1. *Cont.*

Location	Sample	Rock Phase	Th (°C)	Tf (°C)	Tfm (°C)	Tmice (°C)	Salinity%
			106	−79	-	−20.8	22.9
			92	−80	-	23.5	24.7
			90	−81	-	−17.6	20.7
			90	−79	-	−22.1	23.8
			102	−77	-	−22.1	23.8
			92	−74	-	−18.6	21.5
			94	−77	-	−16.9	20.2
			108	−79	-	−22.2	23.8
			108	−80	-	−22.2	23.8
			110	-	-	-	-
			103	-	-	-	-
			102	-	-	-	-
			88	-	-	-	-
			93	-	-	-	-
			92	-	-	-	-
			91	-	-	-	-
			90	-	-	-	-
			85	-	-	−21.6	23.4
			92	-	-	−21.9	23.6
			93	−85	-	-	-
Spelek	Sp. 10	Blocky calcite	Monophase	−39.8		−0.2	0.35
				−37.9		−0.7	1.22
				−39.8		−0.5	0.88
				−37.7		−1.5	2.57
				−39.8		−1.4	2.41
				−39.9		−0.9	1.57
				−39.8		−0.1	0.18
				−37.3		−1.4	2.41
				−39.8		−1.2	2.07
				−41.8		−0.6	1.05
				−39.8		−0.2	0.35
				−40.8		−0.2	0.35

In situ uranium-lead analyses were performed in polished thin sections of dolomite and calcite cements from different samples by laser ablation-inductively coupled plasma-mass spectrometry (LA-ICP-MS) at the Goethe University Frankfurt (Germany). The applied method used an Element2 (Thermo-Scientific, Waltham, MA, USA) sector field ICP-MS coupled to a Resolution ArF excimer laser (Compex Pro 102). The ablated spot size used was 213 μm and depth of the crater was ~20 μm. Samples were screened by LA-ICP-MS for suitable Pb and U concentration and variability. Subsequently, the selected spots were analyzed in fully automated mode overnight. The spot analyses consisted of 20 s background acquisition followed by 20 s of sample ablation. To remove surface contamination prior to analysis, each spot was pre-ablated for 3 s. Soda-lime glass SRM-

NIST 614 was used as a reference glass together with 2 carbonate reference materials, WC-1 and a Zechstein dolomite, to bracket sample analysis (Table 2). The SRM-NIST 614 yielded a depth penetration of about 0.5 $\mu m\ s^{-1}$ and an average sensitivity of 280,000 cps/$\mu g\ g^{-1}$ for ^{238}U. The detection limits for ^{206}Pb and ^{238}U were ~0.1 and 0.05 ng g^{-1}, respectively. All data were corrected by using MS Excel© spreadsheet program [20,21]. The $^{207}Pb/^{206}Pb$ ratio was corrected for mass bias (0.3%) and the $^{206}Pb/^{238}U$ ratio for inter-element fraction (ca. 5%) using SRM-NIST 614. An additional correction of 4% has been applied on the $^{206}Pb/^{238}U$ to correct for difference in the fractionation due to the carbonate matrix. This resulted in a lower intercept age of 23 WC-1 spot analyses of 254.1 ± 1.5 (MSWD = 1.5; anchored at $^{207}Pb/2^{06}Pb$ of 0.851) and 253.9 ± 3.4 (MSWD = 1.5; n = 17) for the Zechstein dolomite, used as in-house reference material in Frankfurt. Data were plotted in the Tera–Wasserburg diagram and ages were calculated as lower intercepts using Isoplot 3.71 [22]. All uncertainties are reported at the 2-sigma level.

Table 2. Concentration of U-Pb and U-Th-Pb isotopes of reference standards.

Grain	^{206}Pb [a] (cps)	U [b] (ppm)	Pb [b] (ppm)	Th [b] / U	^{238}U [c] / ^{206}Pb	±2 s (%)	^{207}Pb [c] / ^{206}Pb	±2 s (%)
NIST-SRM 614								
NIST614 01	151223	0.81	2.37	0.96	1.241	1.8	0.8714	0.6
NIST614 02	152109	0.82	2.40	0.97	1.230	1.9	0.8711	0.6
Nist614 33	153445	0.81	2.37	0.93	1.240	1.8	0.8662	0.5
Nist614 34	150955	0.81	2.36	0.92	1.244	1.8	0.8730	0.6
Nist614 69	150536	0.81	2.36	0.93	1.246	1.9	0.8700	0.6
Nist614 70	154694	0.82	2.43	0.93	1.226	1.9	0.8705	0.5
Nist614 102	148233	0.80	2.34	0.94	1.237	1.8	0.8696	0.5
Nist614 103	149396	0.81	2.36	0.90	1.238	1.8	0.8714	0.6
Nist614 146	146500	0.80	2.33	0.96	1.240	1.9	0.8728	0.6
Nist614 147	149092	0.81	2.36	0.91	1.241	1.8	0.8693	0.5
Nist614 194	150754	0.82	2.42	0.91	1.234	1.8	0.8734	0.6
Nist614 193	144523	0.81	2.35	0.95	1.247	1.8	0.8766	0.6
Nist614 238	146867	0.81	2.39	0.88	1.232	1.8	0.8709	0.5
Nist614 237	147231	0.82	2.39	0.91	1.230	1.8	0.8651	0.6
NIST614 283	145071	0.82	2.40	0.90	1.237	1.8	0.8670	0.6
Nist614 282	144885	0.82	2.40	0.90	1.236	1.8	0.8719	0.6
NISt614 328	146642	0.84	2.46	0.87	1.240	1.9	0.8700	0.5
Nist614 327	141742	0.83	2.41	0.96	1.244	1.8	0.8751	0.6
NIST614 383	142172	0.84	2.45	0.90	1.243	1.8	0.8685	0.6
Nist614 382	141383	0.83	2.43	0.90	1.235	1.8	0.8713	0.6
NIST614 450	143098	0.86	2.51	0.92	1.235	1.8	0.8730	0.5
WC-1								
Calcite 04	57958	3.60	0.24	<0.001	21.30	2.7	0.1674	4.4
Calcite 03	57442	3.58	0.23	<0.001	21.11	2.1	0.1610	3.2
Calcite 30	53037	3.26	0.22	<0.001	20.98	3.0	0.1761	3.1
Calcite 35	55525	3.55	0.22	<0.001	21.22	2.3	0.1519	3.9
Calcite 36	52116	3.08	0.23	<0.001	20.15	4.0	0.1999	10.8
Calcite 67	60273	3.87	0.24	<0.001	21.96	2.3	0.1595	5.3
Calcite 68	50337	2.99	0.22	<0.001	19.87	3.3	0.2024	9.1
Calcite 102	61659	3.57	0.28	<0.001	19.54	2.7	0.2120	6.8
Calcite 103	55004	3.42	0.23	<0.001	20.82	2.3	0.1874	5.1
Calcite 146	53194	3.33	0.24	<0.001	20.20	4.5	0.1916	6.3
Calcite 147	47066	3.01	0.19	<0.001	21.37	2.5	0.1567	4.9
Calcite 195	53947	3.66	0.23	<0.001	21.43	2.6	0.1489	3.7
Calcite 196	52916	3.60	0.21	<0.001	22.27	2.3	0.1489	3.0
Calcite 239	54507	3.76	0.22	<0.001	22.09	2.3	0.1459	2.8
Calcite 240	56367	3.70	0.24	<0.001	21.39	2.2	0.1842	4.3
Calcite 329	53738	3.63	0.23	<0.001	21.39	2.5	0.1598	6.0

Table 2. *Cont.*

Grain	^{206}Pb [a] (cps)	U [b] (ppm)	Pb [b] (ppm)	Th [b] / U	^{238}U [c] / ^{206}Pb	±2 s (%)	^{207}Pb [c] / ^{206}Pb	±2 s (%)
Calcite 384	50897	3.54	0.23	<0.001	21.21	2.6	0.1747	7.6
Calcite 451	51850	3.55	0.25	<0.001	19.98	4.1	0.1843	10.6
Calcite 285	46761	3.18	0.19	<0.001	22.16	2.2	0.1625	5.2
Calcite 452	52066	3.76	0.24	<0.001	21.40	2.4	0.1612	3.2
Calcitee 507	49116	3.34	0.24	<0.001	20.37	2.5	0.2071	6.5
Calcitee 561	49407	3.65	0.23	<0.001	21.36	2.1	0.1661	4.2
Zechstein dolomite								
ZD 05	383976	2.17	3.12	0.02	2.27	5.5	0.7410	0.6
ZD 37	434820	1.59	3.88	0.02	1.35	4.4	0.7685	0.6
ZD 71	145036	2.03	1.15	0.02	5.22	4.2	0.6482	0.8
ZD 106	98172	1.87	0.82	0.03	6.51	3.2	0.6210	0.9
ZD 150	103377	0.88	0.99	0.05	2.89	2.6	0.7279	0.7
ZD 198	38804	0.90	0.36	0.07	6.60	2.4	0.6067	0.7
ZD 242	15608	0.07	0.16	0.12	1.54	4.2	0.7679	0.9
ZD 287	159760	0.86	1.80	0.08	1.64	8.4	0.7706	0.8
ZD 387	104283	4.59	0.80	0.01	13.05	2.9	0.4131	0.6
ZD 197	90226	2.21	0.72	0.02	8.17	3.9	0.5666	1.2
ZD 241	72961	2.16	0.53	0.02	10.03	3.9	0.4992	1.4
ZD 286	57874	2.37	0.37	0.02	13.71	5.2	0.3945	2.8
ZD 331	66408	2.24	0.49	0.00	10.76	2.4	0.4842	0.9
ZD 386	138406	2.21	1.21	0.02	5.72	3.0	0.6317	0.6
ZD 508	114549	2.26	1.08	0.02	5.96	7.7	0.6264	1.3
ZD 562	99812	2.70	0.90	0.02	8.06	3.5	0.5751	1.3
ZD 385	51470	3.65	0.21	0.00	22.01	2.3	0.1443	3.0

Spot size = 213 µm. $^{238}U/^{206}Pb$ error is the quadratic additions of the within run precision (2 SE) and excess of variance of 2% (2 SD). $^{207}Pb/^{206}Pb$ uncertainty propagation as described in Gerdes. Spot size = 213 µm. $^{238}U/^{206}Pb$ error is the quadratic additions of the within run precision (2 SE) and excess of variance of 2% (2 SD). $^{207}Pb/^{206}Pb$ uncertainty propagation as described in Gerdes and Zeh (2009). [a]. Within run background-corrected mean ^{206}Pb signal in cps (counts per second). [b]. U and Pb content and Th/U ratio were calculated relative to NIST SRM614 reference. [c]. corrected for background, within-run Pb/U fractionation (in case of $^{206}Pb/^{238}U$) and subsequently normal-ized to NIST SRM 614 and carbonate matrix effect of 4% for $^{206}Pb/^{238}U$.

4. Results

4.1. Petrography, Cathodoluminescence, SEM, and Stable Isotopes ($\delta^{18}O$ and $\delta^{13}C$)

The Bekhme formation along the Harir-Safin anticlines (NE-Iraq) experienced extensive fracturing, associated with intense hydrothermal fluid migration resulting in zebra-like structures, hydro-brecciation features, and various geodic structures (Figure 2; Salih et al., 2019). The latter authors recognized three phases of saddle dolomites (SD1, SD2, SD3), two phases of blocky calcite cements, and at least three other types of non-saddle dolomites (DI, DII, DIII) inside fractures and geodes.

The DI, DII, and DIII have different sizes and shapes, ranging from anhedral "DI" to euhedral "DIII" crystals (Figure 3A–H), with micritic residues in the core of "DII" and a euhedral shape and transparent color in "DIII" [8]. DIII has a similar dull red luminescence and texture as DI and DII. The micritic residues in "DII" show a spotted red luminescence. The saddle dolomites (SD2 and SD3) display different luminescence patterns ranging from a red luminescence in the cores to a bright red luminescence along the crystal cortices. The blocky calcite crystals, also reported as a hydrothermal product similar to the SD dolomites, have different characteristics, and cathodoluminescence petrography revealed a uniform texture, a dull orange to red luminescence [8].

Calcretized carbonate rocks were a widespread diagenetic feature in the lower and the middle most parts of the Bekhme formation. At least two pedogenic levels were identified based on the presence of typical calcrete features [8]. The vertical variability of pedogenic levels highlights a repeated occurrence of 2–6 m thick calcrete profiles interbedded within a 15–22 m thick succession belonging to the Bekhme formation.

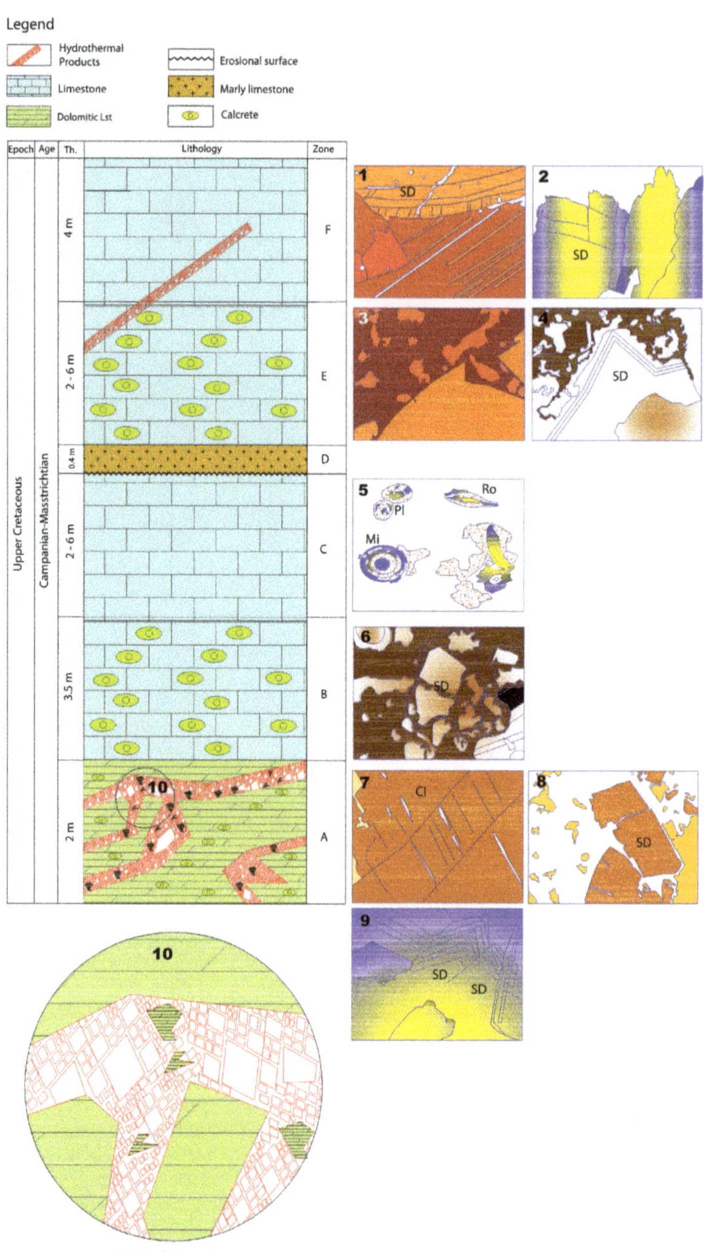

Figure 2. Litho-log of the Bekhme formation from the Spelek-Sulauk area, and next to each zone the digitized microphotographs illustrating the significant alteration of the carbonate rocks. The log shows the emergence (zones B and E) and submergence (zones C and F) of the host carbonate rocks. At least two generations of HT fluids paved their ways along the weakness zones of the host rocks (zones A, E, F). Note the enlarged part (No. 10) of the first generation of HT fluid (zone A), the so-called breccia illustrating the floated, angular grains of the matrix dolomite that is cemented by saddle dolomite. Note another type of brecciation by pedogenic calcrete that caused an in situ fragmentation

of the saddle and matrix dolomites (No. 3, 6, 8). The HT cementation happened prior to the strong in situ brecciation of saddle dolomites (No. 3, 4, 8). Photomicrographs (No. 1, 2, 7, 9) are illustrating unaltered hydrothermal products, while photo No. 4 shows alveolar texture (the brown color on the left part) along the weak joints of the saddle dolomite crystals. The process of alveolar formation is illustrated in photo No. 6; the blue layers define successive layers of dark micrite around fragmented dolomite and show that a non-sequential process of coating starts with an individual tube, followed by a combination of two tubes, until several adjacent tubes are bundled in a micrite coating giving a quasi-concentric structure. CI = blocky calcite, SD = saddle dolomites, Mi = Miliolid, Ro = Rotalid, Pl = Planktonic foraminifera (after [8]).

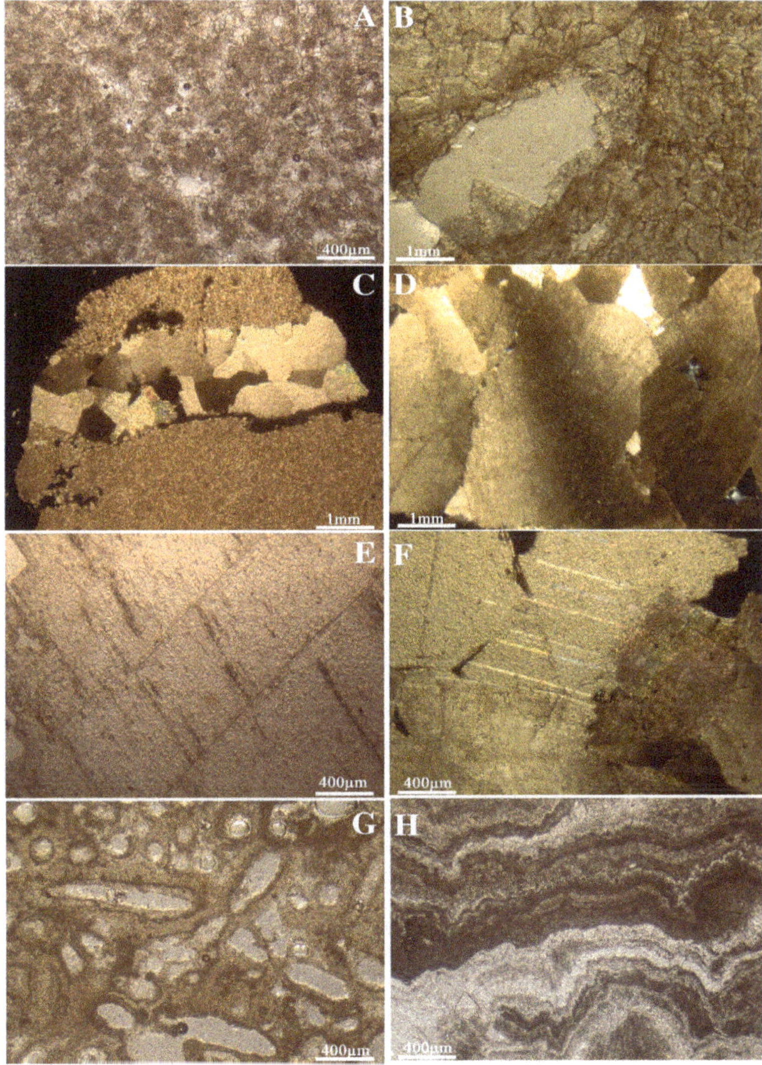

Figure 3. Photomicrographs of carbonate phases. (**A**,**B**) Thin section, polarized light illustrates DI, DII, and DIII dolomites. (**C**,**D**) Thin section under crossed nicols represented the saddle dolomites filling the fracture SD1 (**C**) and void spaces SD2 (**D**). (**E**,**F**) Blocky calcites under PPL (**E**) and XPL (**F**). Calcrete products show two kinds of features: an alveolar texture (**G**) and a pisolitic texture (**H**).

The in situ brecciation due to microbial diagenesis led mostly to alveolar textures, *Microcodium*, and a microbial alteration of the original substrate. A wide spectrum of diagenetic features, such as dissolution, micritization, cementation and re-cementation, dissolution and reprecipitation replacement, grain-to-grain solution, and wall-bridging and open-space fillings have been described [8]. All these features support an in situ pedogenic alteration during subaerial exposure (Figure 3G,H).

All the C-O values used in this study were derived from published data from the same formation and locations [7,8]. All published values from the latter authors have been integrated in this study in order to constrain diagenetic signals based on different recent approaches. The $\delta^{18}O_{VPDB}$ and $\delta^{13}C_{VPDB}$ isotopes gave a narrow range for host limestone and recorded a heavy isotopic composition (varying between −6.7‰ and −4.1‰ for oxygen and 0.3‰ to 1.7‰ for carbon) compared to diagenetic phases (ranging between −17.5‰ and −6.1‰ for oxygen, −6.9‰ and +2.5‰ for carbon). The $\delta^{18}O_{VPDB}$ and $\delta^{13}C_{VPDB}$ values of the calcrete and calcretized samples ranged between −17.8‰ and −6.5‰, and −9.6‰ and 2.5‰, respectively [7,8].

Fragmentation of saddle dolomites floating in the alveolar texture was the most significant feature recognized in the altered saddle dolomites and blocky calcites, and indicates calcretization post-dated saddle dolomite precipitation.

4.2. Petrography of Fluid Inclusions

The study of fluid inclusions (FIs) was carried out on both saddle dolomite and blocky calcite cements (Table 1). Based on the presence, occurrence, and size of the fluid inclusion suitable cements were (i) saddle dolomites SD2 and (ii) SD3, and (iii) blocky calcites CI.

Fluid inclusions in the dolomite and calcite are ubiquitous and included monophase (liquid–L), bi-phase (liquid-vapor, L-V), and sometimes 3-phase (liquid-liquid-vapor L-L-V) inclusions. Two types of FIs were observed in saddle dolomite: primary FIs which were trapped during crystal growth (Figure 4A–C) and secondary FIs which formed after crystal growth (Figures 4A,B and 5A,B). FIs of primary origin occurred in growth zones or in clusters in the crystals (Figure 5C), such as in the core of the dolomite rhombs. Secondary inclusions occurred along trails or in cleavage planes, were frequently larger (up to 40 μm) compared to the primary FIs (up to 20 μm). Due to their small sizes and the darkness of the dolomite crystals, it was very difficult to measure fluid inclusions in the dolomites. The volume of vapor phase varied between 15–25% within the two-phase aqueous inclusions.

Blocky calcite CI was more transparent in comparison with the saddle dolomites. Secondary fluid inclusions were abundant and easily recognized in these blocky calcites. FIs displayed a wider range in size than the dolomite inclusions (measured inclusions range between 15 μm up to >100 μm; Figure 5). The aqueous inclusions (Figure 4A,B) frequently occurred in trails of variable direction, and most of the studied aqueous fluid inclusions were mono-phase. The inclusions displayed a wide range of shapes, from ellipsoidal, simple to irregular. In our study, we measured both primary in saddle dolomites and large secondary monophase FIs in blocky calcite. Since the mono-phase fluid inclusions have no homogenization temperature that can be measured, the final melting temperature of ice and thus the salinity of the inclusions can be measured after artificial stretching of the inclusions.

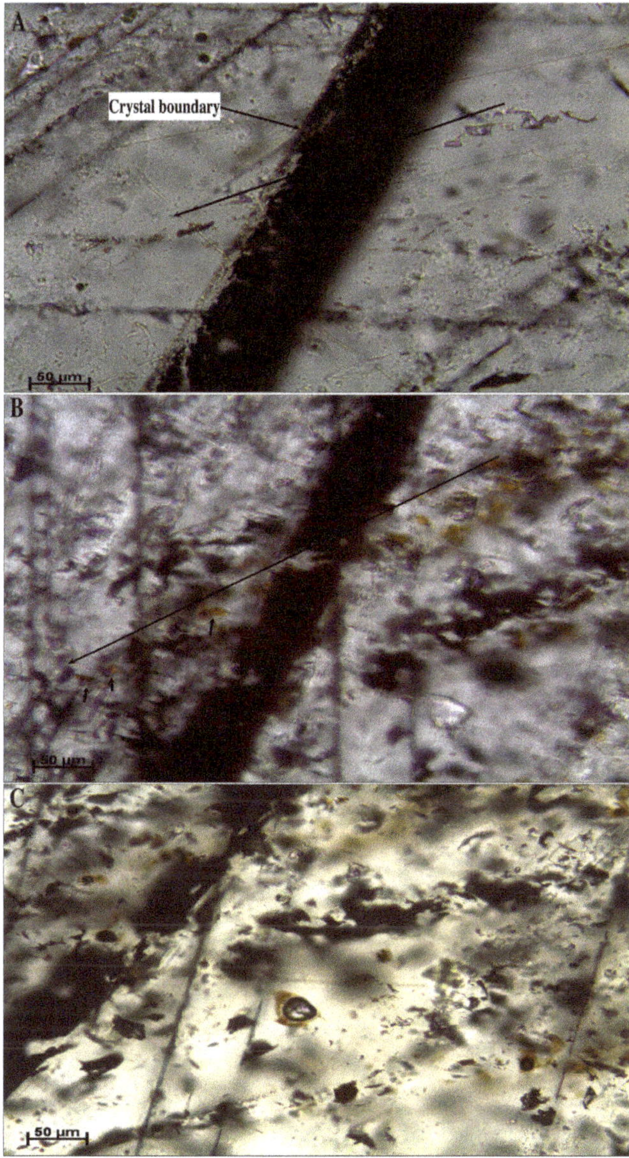

Figure 4. Photomicrographs showing aqueous FIs within the blocky calcite: (**A**) Secondary aqueous FIs crossing the crystal boundary of the blocky calcites (arrow) indicate these inclusions were entrapped after the calcite crystal growth, in addition to small-sized primary FIs. Sua.7. Sulauk section. (**B**) Secondary fluid inclusions crossing the boundary of a calcite crystal, close up the upper right part of the photo that shows the primary FIS. Sulauk section. (**C**) Two-phase primary fluid inclusion. Sua.7. Sulauk section.

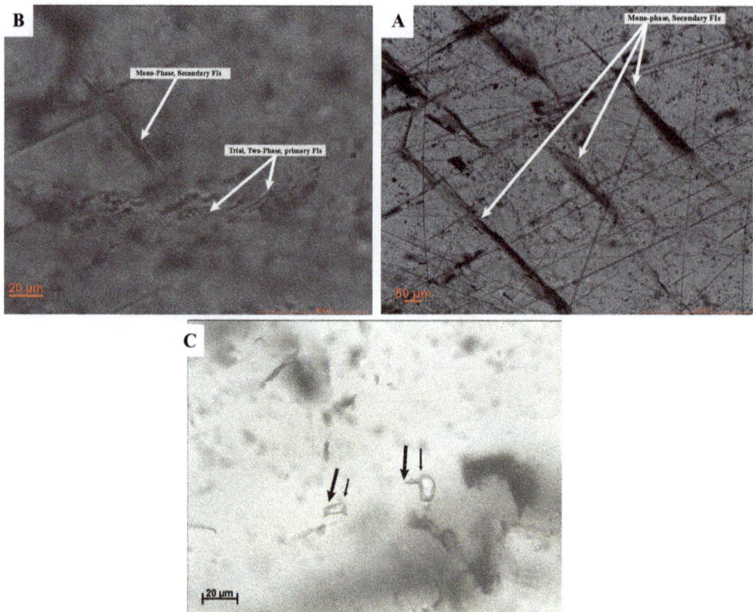

Figure 5. Photomicrographs illustrating one- and two-phase inclusions in the Spelek-Sulauk areas. (**A**,**B**) Secondary, late fluid inclusions in the blocky calcite. (**C**) Primary one-phase and two-phase fluid inclusions (see the black arrows). Sp.10. Spelek section.

4.3. *Fluid Inclusion Microthermometry*

4.3.1. Saddle Dolomite SD2

Homogenization temperatures (Th) of the saddle dolomite-hosted aqueous fluid inclusions ranged between 86 °C and 118 °C (Table 1; Figure 6A). The first melting temperature Tfm ranged between −48.6 °C to −45.7 °C. These values were lower than those for an H_2O-NaCl binary system, and were only slightly higher than the eutectic temperature of ternary H_2O-NaCl-$CaCl_2$ system [23,24]. Ice melting temperatures (Tm_{ice}) of SD2 range between −23.5 °C and −19.3 °C, corresponding to a salinity between 15.9 and 24.7 equivalent weight percent NaCl (eq. wt.% NaCl) (Figure 6). The salinity values were calculated using the equation of Bodnar [25].

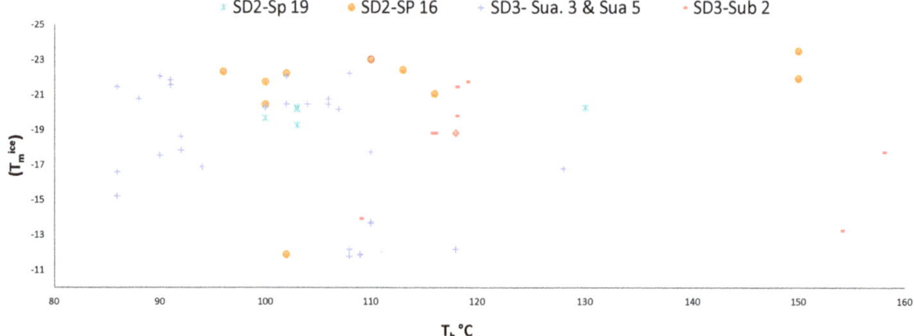

Figure 6. *Cont.*

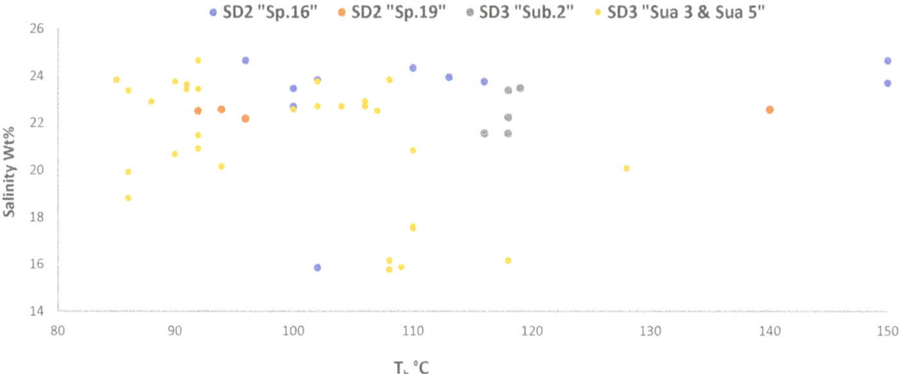

Figure 6. Distribution of homogenization temperatures of the fluid inclusions in the Bekhme formation vs. (**A**) ice melting temperatures (Tmice), and (**B**) corresponding salinities. Sp = Spelek, Sua. and Sub. = Sulauk sections, SD = saddle dolomites.

4.3.2. Blocky Calcites (CI and CII)

CI and CII contained secondary aqueous fluid inclusions. Homogenization temperatures of the FIs in CI were measured on >100 FIs, and showed that generally there are two different types of fluid inclusions, early and late.

Secondary monophase aqueous fluid inclusions were also abundant in blocky calcites. In order to measure their salinities, multiple periods of cooling (up to −100 °C) and heating (up to 200 °C) were carried out to artificially stretch the inclusions and generate a vapor bubble (cf. [26,27]. The monophase secondary FIs yielded a final melting temperature of ice between −0.7 °C to −1.5 °C, corresponding to salinities between 0.2 to 2.6 eq. wt.% NaCl (Table 1). The first melting temperature observed in the fluid inclusions was close to the final melting temperature, which is characteristic for such low salinity inclusions [23].

4.4. U-Pb Dating of the Calcrete

Two samples of calcrete (Sp.3 and Sp.11) were prepared for a LA U-Pb direct dating (Figures 7 and 8; Table 3). Petrographically, sample Sp.3 represents a combination of pure calcrete and chert nodules. Sample Sp.11 displayed an alteration of saddle dolomite inside a geode during the calcretization process. The Pb^{207}/Pb^{206} intercepts at age 3.8 Ma (with σ uncertainties 1.1) and the scatter spot point analyses were outside the isochron with MSWD of 1.9. The pedogenic calcrete and chert nodules from the Spelek sample (Sp.3) yielded a mixing U-Pb age of ~70 Ma for the pedogenic calcrete and ~140 Ma for the chert nodules. Absolute numerical dating results of the same formation of the carbonate cements has been integrated into the recent results that previously reported by Salih [7].

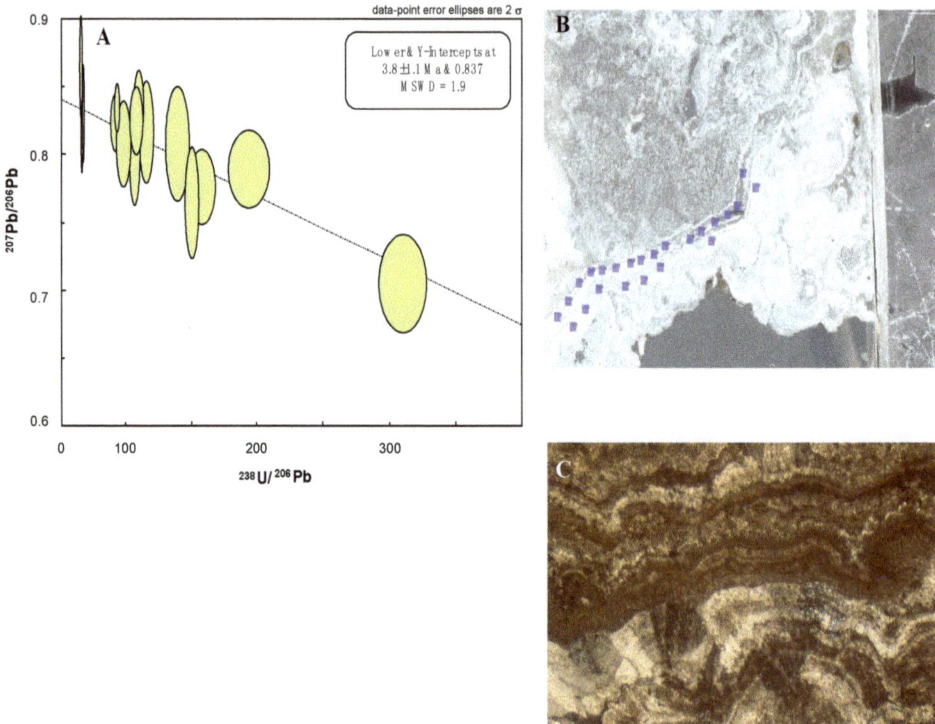

Figure 7. (**A**) The Tera–Wasserburg curve for the analyzed spot points in (**B**). (**B**) Photomicrograph of the ablated spot points inside the geode where saddle dolomite is altered by calcretization process into laminar crust and pisolitic texture. (**C**) The same sample showing the optical characteristics of calcrete features. Sample Sp.11. Spelek section. (n = 24). Harir anticline.

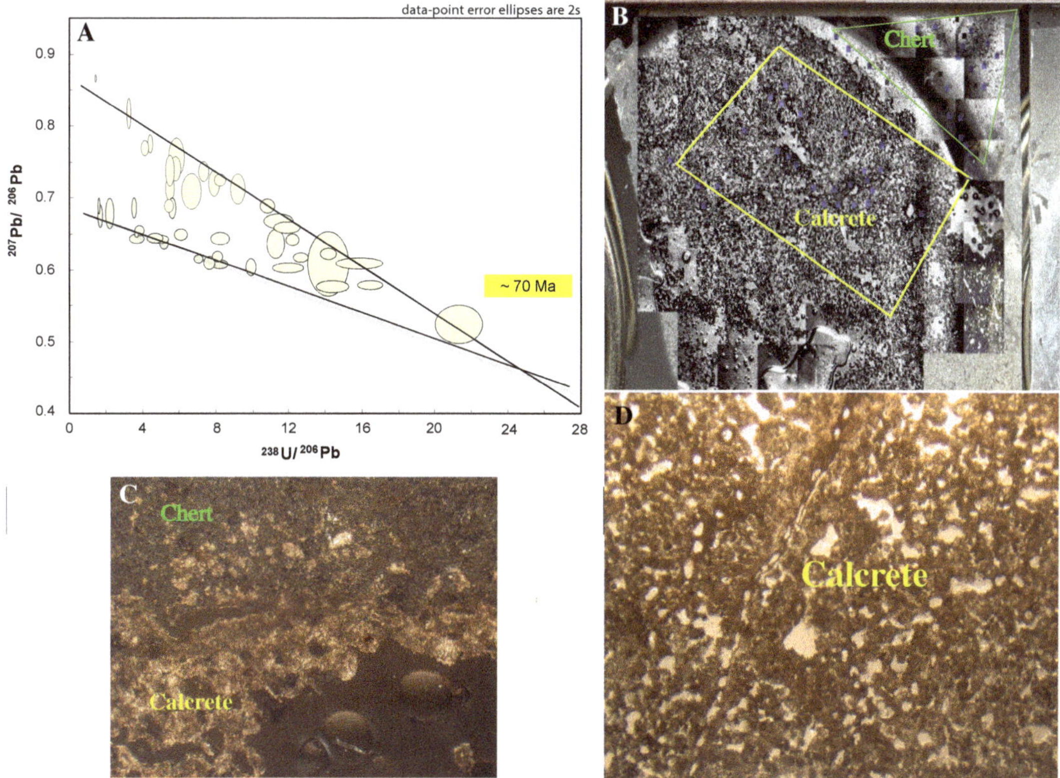

Figure 8. (**A**) The Tera–Wasserburg curve for the analyzed spot points in (**B**), a mixing of two age components in both chert and calcrete is suggested, scatter in the data do not allow to well constrain the age of the two events. (**B**) Photomicrograph showing two subsamples (calcrete and chert), one is an alveolar texture formed as a result of pedogenic calcrete product on the lower part and the center part of the photo (see (**D**)), another subsample located on the upper right part, which represents the location of the chert nodules spot points analysis (see (**C**)). n = 45.

Table 3. Concentration of U-Pb and U-Th-Pb isotopes of host limestone, early dolomite matrix, saddle dolomites, blocky calcites, calcrete, and chert nodules sample by small-scale isochron (SSI) using LAICP-MS approach. Spelek-Sulauk areas. NE-Iraq-Kurdistan region. (n = 190).

Grain	^{206}Pb [a]	U [b]	Pb [b]	Th [b]	^{238}U [d]	±2 s	^{207}Pb [d]	±2 s
	(cps)	(ppm)	(ppm)	U	^{206}Pb	(%)	^{206}Pb	(%)
A172	2196	0.053	0.016	0.006	11.22	3.8	0.6343	2.7
A173	3861	0.056	0.032	0.004	5.425	3.8	0.712	2
A174	4840	0.081	0.038	0.002	6.575	6.6	0.7082	2.9
A175	24795	0.144	0.205	0.008	2.13	8.0	0.6771	2.5
A176	9157	0.086	0.074	0.002	3.473	3.0	0.6851	1.7
A177	7469	0.031	0.060	0.003	1.589	6.9	0.6732	2
A178	6009	0.119	0.052	0.003	7.261	3.0	0.7361	1.6
A179	18528	0.076	0.152	0.003	1.532	3.5	0.6877	1.3
A180	8362	0.125	0.068	0.004	5.538	2.9	0.6846	1.7
A181	4123	0.063	0.035	0.004	5.746	5.8	0.7558	2.9
A182	59898	0.116	0.120	0.005	3.603	9.1	0.6416	0.81
A183	1729	0.102	0.012	0.004	21.29	5.1	0.5217	4.2
A184	5994	0.126	0.051	0.002	7.834	3.0	0.7192	2.1

Table 3. *Cont.*

Grain	^{206}Pb [a] (cps)	U [b] (ppm)	Pb [b] (ppm)	Th [b] / U	^{238}U [d] / ^{206}Pb	±2 s (%)	^{207}Pb [d] / ^{206}Pb	±2 s (%)
A183	1729	0.102	0.012	0.004	21.29	5.1	0.5217	4.2
A184	5994	0.126	0.051	0.002	7.834	3.0	0.7192	2.1
A185	6869	0.082	0.062	0.003	4.342	2.5	0.7749	1.4
A186	3399	0.084	0.029	0.006	9.108	3.6	0.7117	2.5
A187	3879	0.063	0.036	0.005	5.704	4.0	0.7397	1.9
A188	3489	0.040	0.027	0.009	5.394	3.8	0.7375	2.4
A189	37434	1.660	0.283	0.005	16.41	3.5	0.5762	0.94
A190	48897	1.183	0.408	0.005	8.138	4.5	0.6056	0.83
A191	38268	1.672	0.312	0.004	14.32	5.0	0.5741	1.2
A192	39296	1.507	0.325	0.004	12.96	2.7	0.6087	0.82
A193	144628	0.569	1.540	0.068	1.337	2.0	0.8671	0.48
A199	108990	1.226	0.956	0.011	3.767	5.5	0.6533	0.88
A200	21083	0.255	0.208	0.013	4.048	4.0	0.769	1.1
A201	44634	0.670	0.381	0.014	5.072	3.7	0.6346	1
A202	41916	0.659	0.376	0.008	5.342	3.8	0.6876	1.1
A203	35337	0.810	0.298	0.013	7.576	3.3	0.6075	1.2
A204	46046	0.826	0.404	0.008	5.988	4.8	0.6476	0.99
A205	34360	1.100	0.312	0.005	10.75	3.3	0.6875	1.1
A206	64312	0.934	0.547	0.005	4.934	7.0	0.6407	0.73
A207	58540	1.407	0.498	0.016	8.028	2.8	0.6161	1
A208	35535	1.485	0.299	0.004	14.12	2.6	0.6202	1.1
A209	43572	1.063	0.376	0.005	8.147	5.0	0.6417	1.1
A210	70359	1.446	0.598	0.007	6.963	3.0	0.6131	0.77
A211	48935	1.820	0.407	0.004	12.59	2.6	0.6147	0.86
A212	40903	1.514	0.349	0.004	11.88	6.0	0.6005	0.94
A213	42144	2.265	0.526	0.014	9.835	2.6	0.6014	1.6
A214	70012	0.943	0.594	0.010	4.613	7.7	0.6425	0.85
A215	39667	1.337	0.446	0.004	8.152	3.1	0.7238	0.96
A216	40872	1.952	0.346	0.005	15.83	6.7	0.6067	1
A217	41833	1.414	0.372	0.004	11.36	6.0	0.6665	1.3
A218	36369	1.297	0.323	0.005	11.84	4.9	0.6574	1
A219	33775	1.219	0.292	0.007	12.15	2.4	0.6402	1.1
A220	257672	2.773	2.255	0.006	3.604	2.5	0.6476	0.47
A221	10013	0.094	0.104	0.491	3.166	2.8	0.8176	2.2

5. Interpretation and Discussion

5.1. Fluid Inclusion Petrography and Composition of Fluid Flow

Petrographical observations indicate that saddle dolomites precipitated episodically during an early and late paragenetic sequence, related to subaerial condition and submersion [8]. Saddle dolomites precipitated in two times or phases: Late Cretaceous and Tertiary, the absolute ages for these two phases have been measured by LA U-Pb isotopes and gave numerical absolute ages of 73.8 Ma, and 14.5 and 8.6 Ma [7].

The fluid inclusion petrography shows that most of the measured FIs in the saddle dolomites were aqueous fluid inclusion FIAs. The primary and secondary FIAs were the most dominant types, therefore the FIAs were entrapped during and after the crystal's growth. Therefore, fluid inclusion petrography in saddle dolomites reveals two episodes of entrapments, and thus more than one phase of HT fluid was channeled through the Upper Cretaceous Bekhme formation. However, these two episodes happened at two different times, but still the large-sized secondary (mono-phase) fluid inclusions were trapped at the final stage of diagenesis, since the latter FIs lack the vapor phase and were characterized by large secondary FIs.

The homogenization temperature values from primary and secondary FIAs largely overlapped. The data from both primary and secondary inclusions show that entrapment temperatures of these FIs occurred in two populations (Figures 6 and 9): a first major population of Th values ranging between 83 °C and 120 °C, and a second minor population with Th values between 130 °C and 160 °C. Lining of fractures and geodes by a first generation of saddle dolomite followed by a second generation of saddle dolomite are consistent with high temperature fluids. At high temperature, an intensive fluid-rock interaction is expected, and such an intense interaction has been reported previously by Salih [3,8] utilizing O-C isotopes. The strongly negative oxygen isotope values (as low as −18‰) and the wide range of O-C values indicate the involvement of several episodes of HT fluid flow through the Bekhme carbonate sequence (Figures 10 and 11). This interaction caused the precipitation of different types of saddle dolomites (SD1, SD2, and SD3) (Figure 3). However, the Th values in saddle dolomites SD2 and SD3 are overlapping (Figure 9). The lowest homogenization temperature (86 °C), however, concerned the earliest precipitated saddle dolomite, while the highest temperature of entrapment was present in a late saddle dolomite (160 °C; Figure 9).

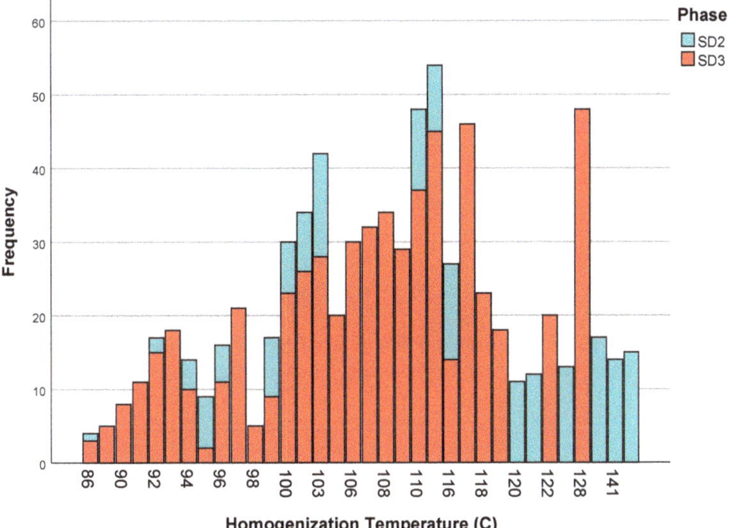

Figure 9. Th data from primary and secondary fluid inclusions in the saddle dolomites and blocky calcite cements vs. frequency from the Spelek-Sulauk area. Each color represents data from different diagenetic phases (n = 102).

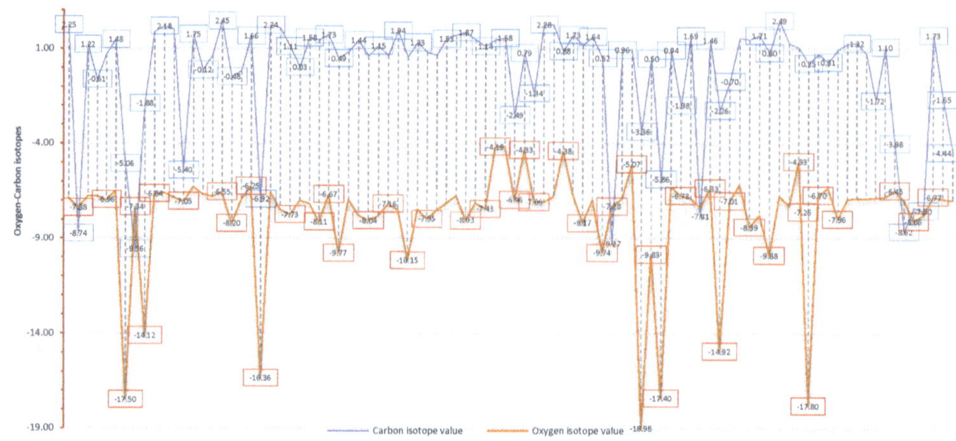

Figure 10. Oxygen and carbon isotopic compositions of carbonate rocks of the Bekhme formation and diagenetic phases for each generation of dolomites, calcites, and calcrete deposits (modified after Salih [7,8]).

The highest cluster of salinity (to 25 eq. wt.% NaCl) is seven times the salinity of seawater. Therefore, the flux of these fluids suggests that they originated as sedimentary brines migrating from the deeper subsurface into the Bekhme formation along deep-seated faults. The upward migration of hydrothermal fluids in the subsurface is often related to a tectonic regime that actively pumped fluids repeatedly during the reactivation of fault systems [7]. The intensive fracturing and the hydraulic breccias in the study area could confirm the involvement of a tectonic regime causing the recorded increase of the Th values in the FIs analyses of the saddle dolomites. The cement's filling open spaces were previously documented as an extra-light value of $\delta^{18}O$ and $\delta^{13}C$ and reported that the fluids involved in precipitation of these cements were hydrothermal fluids (see Figures 10 and 11). Therefore, the following observation and data are considered as indicative of intensive and episodic reactivation of tectonic movement and significant variation was present in the composition of hydrothermal fluids: (i) precipitation of multiple saddle dolomites; (ii) fluctuation of homogenization temperatures (e.g., wide range of paleotemperatures) in FIs trapped in the saddle dolomites, and (iii) saddle dolomites evidencing a consistent fluid origin; (iv) the intensive and moderate interaction with different lithological units in the sequence (based on stable isotope data) and the wide distribution range of oxygen-carbon isotopes (Figure 11).

Recently, a radiometric study also provided a precise time frame using micro-laser ablation technique by ICP-MS on dolomites and calcites hosted in the Upper Cretaceous carbonates [7]. The authors identified more than one type of hydrothermal fluids that altered the Upper Cretaceous Bekhme carbonates and gave an absolute numerical age within the Maastrichtian and Tertiary time interval. The occurrence of a complex fracture network, hydrothermal breccia, geodes, and other open spaces are generally related to more than one faulting system in the study area [8].

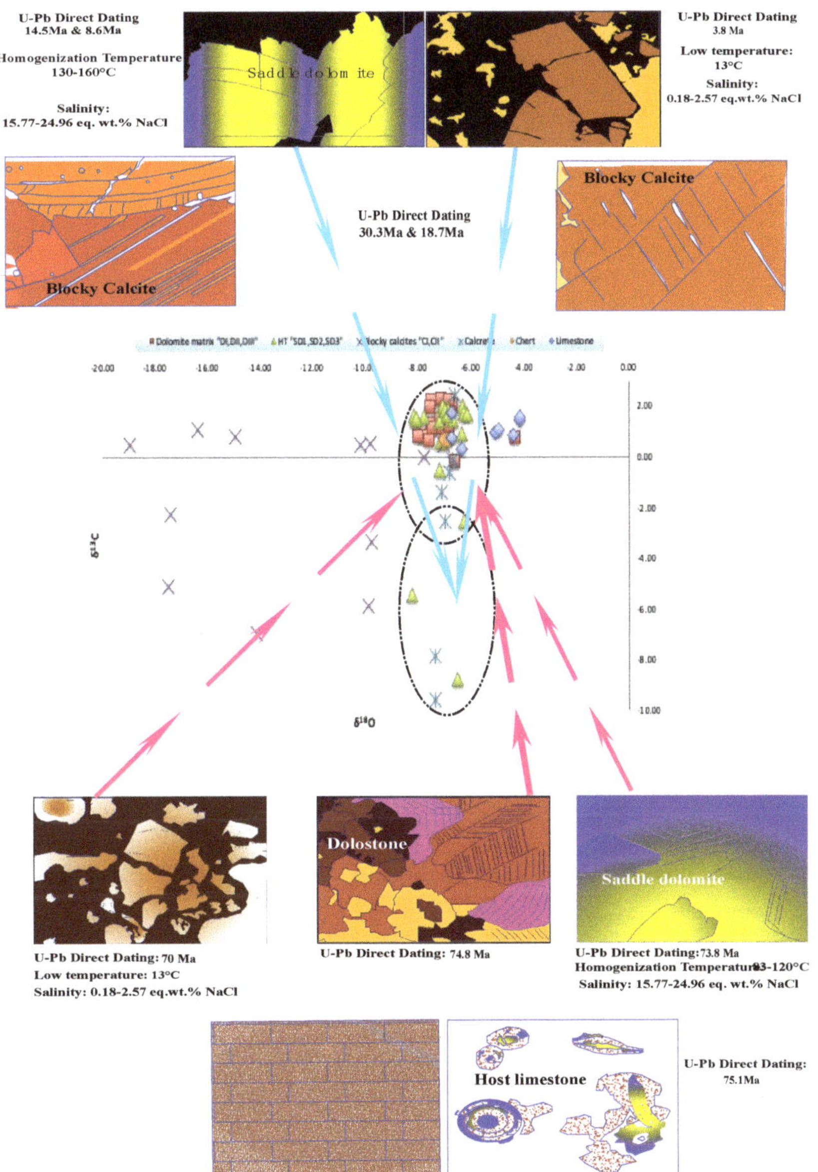

Figure 11. Radiometric model integrated with petrographic and geochemical data to illustrate the absolute age and chemistry of dolomitizing fluids and host rocks during and after the depositional time.

5.2. Thermal Re-Equilibrium, Paleotemperatures in Repeated Hydrothermal Fluids

Fluid inclusion microthermometry, and stable isotopes data provide detailed information on the evolution of temperature and fluid compositions of the HT carbonates, matrix dolomite, and calcrete products. The homogenization temperatures from saddle dolomite samples cover a considerable range of values. More than 95% of all measured primary and secondary fluid inclusions point to homogenization temperatures between 83 °C and

118 °C for SD2/SD3. This range represents true syn-entrapping and post-entrapping temperatures during and after mineral growths, either in or out of thermal re-equilibrium.

Chemical and physical factors affecting thermal re-equilibrium processes can produce stretching, leakage, and refilling of the fluid inclusions [24,28]. However, the inability of fluid inclusions to resist a temperature increase of the host rock increases with the size of the inclusions, causing stretching of the inclusions [29]. The current measurements of small-sized inclusions (15–20 μm) with the same Th values as the larger inclusions (up to 40 μm) suggest that "re-equilibrium" of inclusions did not play a major role in our data set.

Multiple events of hydrothermal fluids over time based on bimodal paleotemperature distributions are documented throughout the Permo-Triassic and Carboniferous sections in the West of Britain and Ireland [30]. Consequently, the wide and high-recorded temperatures (83–160 °C) in our study may reflect either multiple pulses of hydrothermal fluids (cf. [31]) or a fluid evolution in the context of a long-lived hydrothermal system [32]. An evolution of the hydrothermal fluids over long periods of time (long-lived HT fluids, cf. [33]) is linked to a considerable range of salinity composition and associated with relatively uniform temperature degrees.

5.3. *Cold Meteoric Waters' Origin Associated with Hydrothermal Products*

High saline fluids and low saline meteoric waters were brought either by deep-seated faults or by exposing the carbonate body to the surface, which resulted in precipitation of different carbonates [8]. The exposure of carbonate to the surface may have brought low saline water into the Bekhme formation, with the trapping of low saline fluid inclusions (0.2 and 2.6 eq. wt.% NaCl). These FIs are trapped as a secondary monophase inclusions in calcite cement, and are indicative of a late episode of diagenesis. These inclusions were entrapped along cleavage planes in calcite, and they crosscut even the boundary of crystal grain (Figure 5). Low saline water involved in the precipitation of this kind of calcite displayed a much lower temperature (around 13 °C, [8]) than the ones of hydrothermal saddle dolomites (83–118 °C). Similar large fluid inclusions were trapped below <50 °C in the Lower Carboniferous of the Variscan foreland of southern Belgium, and these FIs postdated the Variscan orogeny. Fresh water from which these calcites precipitated have a telogenetic origin (e.g., [27,34]). Invariable δ18 O values and variable and negative δ13C values of the calcites studied closely match the inverted J-pattern or curve of meteoric calcites, and therefore a meteoric origin and telogenetic origin is proposed for these late calcites (Figures 10 and 11).

5.4. *Radiogenic Isotopes Data Scanning*

5.4.1. Timing of Diagenetic Events Utilizing U-Pb Dating

Using the absolute laser ablation ICP-MS U-Pb direct dating, the saddle dolomites and blocky calcites originated from hydrothermal fluid and occurred at least in two major phases (Cretaceous 73.8 Ma and Tertiary, not older than 30.8 Ma [7]). These meaningful ages overlap with an estimated age of the studied Upper Cretaceous carbonate (75.1 Ma) (Figure 11). Visual and optical characteristics of carbonates show evidence of microbial alteration on the surface due to the ancient activity of fungal and bacterial colonization [8].

U-Pb dating of calcrete samples "Sp.3 and Sp.11" yielded two ages: ~70 Ma and 3.8 Ma. According to Gradstein [35], these numerical absolute ages are well consistent with the Campanian and Pliocene periods. Calcrete sample Sp.3 with a clear alveolar texture gives a Campanian age of ~70 Ma. The numerical absolute age "~70 Ma" is equivalent to the interval age of the Bekhme depositional carbonates (Campanian-Early Maastrichtian age). The pisolitic and laminar crusts of the calcrete (with a lack of pedogenic products, e.g., alveolar texture) on the altered saddle dolomite yielded a U-Pb age of 3.8 Ma. This age of 3.8 Ma records the last event of the diagenetic history in Upper Cretaceous carbonates.

Based on these two different events, the first one (~70 Ma) is in good agreement with the estimated age of Bekhme formation, while the second one brings new insight, indicating that a young diagenetic phase affected the whole Bekhme formation. This diagenetic event could have coincided with the migration of a low saline fluid (0.2 and 2.6 eq. wt.% NaCl) recorded in the blocky calcites. The uplifting of the area during a major tectonic event in the Pliocene could have caused widespread infiltration of meteoric water, whose salinity increased due to the interaction with the host rock, and mixing with higher salinity fluids, still present in the pores. The Pliocene recorded the greatest rate of erosion in the high folded zone [36].

Combining petrography, O-C isotopic, and U-Pb data, first by tying the two calcrete types with oxygen-carbon isotope compositions, shows that the O-C isotope values of the older calcrete (70 Ma) overlap with those of the saddle dolomites and the matrix dolomites (73.8 Ma and 74.8 Ma), in contrast to the younger calcrete (3.8 Ma) which has lighter O-C isotope values (Figure 11).

5.4.2. Timing of Non-Diagenetic and Diagenetic Carbonate Rocks

Using the U-Pb age dating method and by integrating field observations, petrography, and fluid inclusion microthermometry from carbonate rocks, Salih [8] drew a time framework for the host limestone of the Bekhme formation and its diagenetic alteration. The meaningful numerical age of unaltered host limestone (non-diagenetic or pristine) is 75.1 Ma, and corresponds to a Late Campanian age, in good concordance with the interval age of the Bekhme formation [7]. The authors added that in a very short time, the early dolomites replaced the host limestones (~73.8 Ma). Another phase of an HT mineral has an absolute age of Tertiary times. An early HT mineral just postdated the precipitation of matrix dolomite (~74.8 Ma; [7]). Therefore, saddle dolomites and blocky calcites recorded at least two major generations of HT fluids.

Another new diagenetic phase consists of the pedogenic calcrete, which shows an enrichment in uranium isotopes. This enrichment suggests that the sediment 'or calcretized rock', is younger than the other carbonate phases. The numerical absolute age of one sample of the calcrete associated with abundant alveolar textures and other microbial activities gives an age of ~70 Ma, which is close to the Bekhme formation time and possibly the first generation of the HT dolomites. In the case of the U-Pb dating of the second calcrete level with the laminar and pisolitic textures, and without alveolar textures, an age of 3.8 Ma was established (Figure 11).

6. Conclusions

Based on optical characteristics, visual petrography of fluid inclusions and microthermometry, stable isotopes, and U-Pb dating, two main entrapment episodes of aqueous FIs were identified from their optical characteristics at room temperatures;

1. An early entrapment episode of fluid inclusions in HT dolomite cements exhibited a range of homogenization temperatures with (i) a first population of Th values between 83 °C and 120 °C and (ii) a second population with Th values between 130 °C and 160 °C. Both populations showed an overlap in their salinity (mainly between 15.8–25 eq. wt.% NaCl). Their salinity and Th values indicated an episodic upward migration of HT fluids related to tectonic activity;
2. The late entrapment episode of fluid inclusions was associated with secondary monophasic FIs, and recorded a diluted fluid (up to 2.6 eq. wt.% NaCl) under low temperature conditions, which coincided with a calcretization of the carbonate rock;
3. The early migration of high temperature fluids possibly occurred during the Late Cretaceous HT, while the second migration of low saline fluids of meteoric origin was associated with and uplifted during Pliocene times. These two periods of fluid migration provide new insight into the geology of the area studied, i.e., contributing to the relationship between fluid flow and the tectonic movements during the Zagros Orogeny;
4. The two episodes are further supported by LA ICP-MS U-Pb dating. The numerical results indicated an absolute age of the pedogenic calcrete, associated with abundant alveolar textures and other microbial activities, which gave a U-Pb age of ~70 Ma

which is close to the period that the Bekhme formation was deposited. This paragenetic sequence ended by the uplift of the Spelek-Sulauk areas during the Pliocene, which caused calcrete formation. In the case of the U-Pb age dating of the laminar and pisolitic textures, an age of 3.8 Ma was established for the end of the diagenetic event when this sequence was exposed to the surface.

Author Contributions: Writing and preparation the original draft, N.S. and A.P.; review and editing, N.S., A.P., P.M. and H.M.; Methodology and software, G.A., P.M., A.P. and N.S. All authors have read and agreed to the published version of the manuscript.

Funding: The study benefited from research funds of the Université Libre de Bruxelles (ULB)-Belgium.

Acknowledgments: This work was supported by Université Libre de Bruxelles-Belgium and it is FIERCE contribution No. 86.

Conflicts of Interest: The authors declare no conflict of interest.

References

1. Hardie, L.A. Secular variation in seawater chemistry: An explanation for the coupled secular variation in the mineralogies of marine limestones and potash evaporites over the past 600 m.y. *Geology* **1996**, *24*, 279–283. [CrossRef]
2. Mansurbeg, H.; Morad, D.; Othman, R.; Morad, S.; Ceriani, A.; Al-Aasm, I.; Kolo, K.; Spirov, P.; Proust, J.N.; Preat, A.; et al. Hydrothermal dolomitization of the Bekhme formation (Upper Cretaceous), Zagros Basin, Kurdistan Region of Iraq: Record of oil migration and degradation. *Sediment. Geol.* **2016**, *341*, 147–162. [CrossRef]
3. Salih, N.; Mansurbeg, H.; Préat, A. Geochemical and Dynamic Model of Repeated Hydrothermal Injections in Two Mesozoic Successions, Provençal Domain, Maritime Alps, SE-France. *Minerals* **2020**, *10*, 775. [CrossRef]
4. Yoshimura, T.; Wakaki, S.; Ishikawa, T.; Gamo, T.; Araoka, D.; Ohkouchi, N.; Kawahata, H. A Systematic Assessment of Stable Sr Isotopic Compositions of Vent Fluids in Arc/Back-Arc Hydrothermal Systems: Effects of Host Rock Type, Phase Separation, and Overlying Sediment. *Front. Earth Sci.* **2020**, *8*, 595711. [CrossRef]
5. Chi, G.; Larryn, W.D.; Huanzhang, L.; Jianqing, L.; Haixia, C. Common Problems and Pitfalls in Fluid Inclusion Study: A Review and Discussion. *Minerals* **2021**, *11*, 7. [CrossRef]
6. Farsang, S.; Louvel, M.; Zhao, C.; Mezouar, M.; Rosa, A.D.; Widmer, R.N.; Feng, X.; Liu, J.; Redfern, S.A. Deep carbon cycle constrained by carbonate solubility. *Nat. Commun.* **2021**, *12*, 1–9. [CrossRef]
7. Salih, N.; Mansurbeg, H.; Kolo, K.; Gerdes, A.; Préat, A. In situ U-Pb dating of hydrothermal diagenesis in tectonically controlled fracturing in the Upper Cretaceous Bekhme Formation, Kurdistan Region-Iraq. *Int. Geol. Rev.* **2019**, *62*, 2261–2279. [CrossRef]
8. Salih, N.; Mansurbeg, H.; Kolo, K.; Préat, A. Hydrothermal Carbonate Mineralization, Calcretization, and Microbial Diagenesis Associated with Multiple Sedimentary Phases in the Upper Cretaceous Bekhme Formation, Kurdistan Region-Iraq. *Geosciences* **2019**, *9*, 459. [CrossRef]
9. Katz, D.A.; Eberli, G.P.; Swart, P.K.; Smith, L.B. Tectonic-hydrothermal brecciation associated with calcite precipitation and permeability destruction in Mississippian carbonate reservoirs, Montana and Wyoming. *AAPG Bull.* **2006**, *90*, 1803–1841. [CrossRef]
10. Jack, S.; Cathy, H.; Hilary, C.; Ardiansyah, K. Burial dolomitization driven by modified seawater and basal aquifer-sourced brines: 562 Insights from the Middle and Upper Devonian of the Western Canadian Sedimentary Basin. *Basin Res.* **2020**, *33*, 648–680. [CrossRef]
11. Zhu, D.; Liu, Q.; Zhang, J.; Ding, Q.; He, Z.; Zhang, X. Types of Fluid Alteration and Developing Mechanism of Deep Marine Carbonate Reservoirs. *Geofluids* **2019**, *2019*, 1–18. [CrossRef]
12. Davies, G.R.; Smith, L.B. Structurally controlled hydrothermal dolomite reservoir facies: An overview. *AAPG Bull.* **2006**, *90*, 1641–1690. [CrossRef]
13. Lonnee, J.; Machel, H.G. Pervasive dolomitization with subsequent hydrothermal alteration in the Clarke Lake gas field, Middle Devonian Slave Point Formation, British Columbia, Canada. *AAPG Bull.* **2006**, *90*, 1739–1761. [CrossRef]
14. Bellen, V.; Dunnington, H.; Wetzel, R.; Morton, D. Lexique stratigraphique international Asie. *Iraq* **1959**, *3C*, 333.
15. English, J.; Lunn, G.A.; Ferreira, L.; Yacu, G. Geologic evolution of the Iraqi Zagros, and its influence on the distribution of hydrocarbons in the Kurdistan region. *AAPG Bull.* **2015**, *99*, 231–272. [CrossRef]
16. Ismail, S.; Kettanah, Y.; Chalabi, S.; Ahmed, A.; Arai, S. Petrogenesis and PGE distribution in the Al-and Cr-rich chromites of the Qalander ophiolite, northeastern Iraq: Implications for the tectonic environment of the Iraqi Zagros Suture Zone. *Lithos* **2014**, *202*, 21–36. [CrossRef]
17. Ballato, P.; Uba, C.; Landgraf, A.; Strecker, M.; Sudo, M.; Stockli, D.; Friedrich, A.; Tabatabaei, S. Arabia-Eurasia continental 501 collision: Insights from late Tertiary foreland basin evolution in the Alborz Mountains, northern Iran. *Geol. Soc. Am. Bull.* **2011**, *123*, 106–131. [CrossRef]

18. Homke, S.; Vergés, J.; Serra-Kiel, J.; Bernaola, G.; Sharp, I.; Garcés, M.; Montero-Verdú, I.; Karpuz, R.; Goodarzi, M.H. Late Cretaceous-Paleocene formation of the proto–Zagros foreland basin, Lurestan Province, SW Iran. *Geol. Soc. Am. Bull.* **2009**, *121*, 963–978. [CrossRef]
19. Muchez, P.; Marshall, J.D.; Touret, J.L.R.; Viaene, W.A. Origin and migration of palaeofluids in the Upper Visean of the Campine Basin, northern Belgium. *Sedimentology* **1994**, *41*, 133–145. [CrossRef]
20. Gerdes, A.; Zeh, A. Combined U–Pb and Hf isotope LA-(MC-)ICP-MS analyses of detrital zircons: Comparison with SHRIMP and new constraints for the provenance and age of an American metasediment in central Germany. *Earth Planet. Sci. Lett.* **2006**, *249*, 47–61. [CrossRef]
21. Gerdes, A.; Zeh, A. Zircon formation versus zircon alteration—New insights from combined U–Pb and Lu–Hf in-situ LA-ICP-MS analyses, and consequences for the interpretation of Archean zircon from the Central Zone of the Limpopo Belt. *Chem. Geol.* **2009**, *261*, 230–243. [CrossRef]
22. Ludwig, L. *Isoplot 3.7: A Geochronological Toolkit for Microsoft Excel*; Special Publication 4; Berkeley Geochronology Center: Berkeley, CA, USA, 2008; pp. 1–77.
23. Roedder, E. Fluid inclusions. *Rev. Mineral.* **1984**, *12*, 646.
24. Goldstein, H. Fluid inclusions in sedimentary and diagenetic systems. *Lithos* **2001**, *55*, 159–193. [CrossRef]
25. Bodnar, R. Revised equation and table for determining the freezing point depression of H2O-Nacl solutions. *Geochim. Cosmochim. Acta* **1993**, *57*, 683–684. [CrossRef]
26. Goldstein, R.H. Petrographic and geochemical evidence for origin of paleospeleothems, New Mexico: Implications for the application of fluid inclusions to studies of diagenesis. *J. Sediment. Petrol.* **1990**, *60*, 282–292.
27. Muchez, P.; Slobodnik, M. Recognition and significance of multiple fluid inclusion generations in telogenetic calcites. *Min.-Alogical Mag.* **1996**, *60*, 813–819.
28. Bodnar, J. Reequilibration of fluid inclusions. In *Fluid Inclusions—Analysis and Interpretation*; Samson, I., Anderson, A., Marshall, D., Eds.; Short Course Series; Mineralogical Association of Canada: Quebec City, QC, Canada, 2003; Volume 32, pp. 213–231.
29. Goldstein, H.; Reynolds, J. Systematics of fluid inclusions in diagenetic minerals. *SEPM Soc. Sediment. Geol.* **1994**, *545*, 31.
30. Middleton, J.; Parnell, J.; Green, F.; Xu, G.; McSherry, M. *Hot Fluid Flow Events in Atlantic Margin Basins: An Example from the Rathlin Basin*; Special Publications 188; Geological Society: London, UK, 2001; pp. 91–105.
31. Hiemstra, J.; Goldstein, H. *Repeated Injection of Hydrothermal Fluids into Downdip Carbonates: A Diagenetic and Stratigraphic Mechanism for Localization of Reservoir Porosity, Indian Basin Field, New Mexico, USA*; Special Publications 406; Geological Society: London, UK, 2015; pp. 141–177.
32. Tobin, C.; Claxton, L. Multidisciplinary thermal maturity studies using vitrinite reflectance and fluid inclusion microthermom-601 etry, A new calibration of old techniques. *Am. Assoc. Pet. Geol. Bull.* **2000**, *84*, 1647–1665.
33. Dewit, J.; Foubert, A.; Desouky, E.A.; Muchez, P.; Hunt, D.; Vanhaecke, F.; Swennen, R. Characteristics, genesis and parame-ters controlling the development of a large stratabound HTD body at Matienzo (Ramales Platform, Basque Cantabrian Basin, northern Spain). *Mar. Pet. Geol.* **2014**, *55*, 6–25. [CrossRef]
34. Alonso-Zarza, M.; Tanner, H. *Carbonates in Continental Setting: Facies, Environments and Processes, Developments in Sedimentology*; Elsevier: Amsterdam, The Netherlands, 2010; pp. 225–267.
35. Gradstein, M.; Ogg, G.; Schmitz, M.; Ogg, G. *The Geologic Time Scale 2012*; Elsevier: Boston, MA, USA, 2012; p. 1176.
36. Lawa, F.A.; Koyi, H.; Ibrahim, A. Tectono-stratigraphic evolution of the NW segment of the Zagros Fold-Thrust Belt, Kurdistan, NE. *J. Pet. Geol.* **2012**, *36*, 75–96. [CrossRef]

MDPI

Article

Tracking the Origin and Evolution of Diagenetic Fluids of Upper Jurassic Carbonate Rocks in the Zagros Thrust Fold Belt, NE-Iraq

Namam Salih [1,2,*], Alain Préat [3], Axel Gerdes [4], Kurt Konhauser [5] and Jean-Noël Proust [6]

1 Engineering Faculty, Petroleum Engineering Department, Soran University, Soran-Erbil 44008, Iraq
2 Scientific Research Centre (SRC), Soran University, Soran-Erbil 44008, Iraq
3 Res. Grp.—Biogeochemistry & Modelling of the Earth System, Université Libre de Bruxelles, 1050 Brussels, Belgium; apreat@ulb.be
4 Institute for Geosciences, Goethe-University Frankfurt, Altenhöferallee 1, 60438 Frankfurt am Main, Germany; gerdes@em.uni-frankfurt.de
5 Department of Earth and Atmospheric Sciences, University of Alberta, Edmonton, AB T6G 2E3, Canada; kurtk@ualberta.ca
6 Géosciences, CNRS, Rennes University, 35000 Rennes, France; jean-noel.proust@univ-rennes1.fr
* Correspondence: namam.salih@soran.edu.iq

Citation: Salih, N.; Préat, A.; Gerdes, A.; Konhauser, K.; Proust, J.-N. Tracking the Origin and Evolution of Diagenetic Fluids of Upper Jurassic Carbonate Rocks in the Zagros Thrust Fold Belt, NE-Iraq. *Water* **2021**, *13*, 3284. https://doi.org/10.3390/w13223284

Academic Editor: Domenico Cicchella

Received: 18 October 2021
Accepted: 12 November 2021
Published: 19 November 2021

Publisher's Note: MDPI stays neutral with regard to jurisdictional claims in published maps and institutional affiliations.

Abstract: Utilizing sophisticated tools in carbonate rocks is crucial to interpretating the origin and evolution of diagenetic fluids from the Upper Jurassic carbonate rocks along the Zagros thrust-fold Belt. The origin and evolution of the paleofluids utilizing in-situ strontium isotope ratios by high resolution laser ablation ICP-MS, integrated with stable isotopes, petrography and fieldwork are constrained. Due to the lack of information on the origin of the chemistry of the fluids, the cements that filled the Jurassic carbonate rocks were analysed from the fractures and pores. This allowed us to trace the origin of fluids along a diagenetic sequence, which is defined at the beginning from the sediment deposition (pristine facies). Based on petrography and geochemistry (oxygen-, carbon- and strontium-isotope compositions) two major diagenetic stages involving the fluids were identified. The initial stage, characterized by negative $\delta^{13}C_{VPDB}$ values (reaching −10.67‰), involved evaporated seawater deposited with the sediments, mixed with the input of freshwater. The second stage involved a mixture of meteoric water and hot fluids that precipitated as late diagenetic cements. The late diagenetic cements have higher depleted O–C isotope compositions compared to seawater. The diagenetic cements display a positive covariance and were associated with extra- $\delta^{13}C_{VPDB}$ and $\delta^{18}O_{VPDB}$ values (−12.87‰ to −0.82‰ for $\delta^{18}O_{VPDB}$ and −11.66‰ to −1.40‰ for $\delta^{13}C_{VPDB}$ respectively). The distinction between seawater and the secondary fluids is also evident in the $^{87}Sr/^{86}Sr$ of the host limestone versus cements. The limestones have $^{87}Sr/^{86}Sr$ up to 0.72859, indicative of riverine input, while the cements have $^{87}Sr/^{86}Sr$ of (0.70772), indicative of hot fluid circulation interacting with meteoric water during late diagenesis.

Keywords: origin of diagenetic fluids; strontium isotope-laser ablation ICP-MS; upper jurassic carbonate rocks; ZTFB; NE-Iraq

1. Introduction

The Upper Jurassic Barsarin formation is located along the Zagros thrust-fold belt (ZTFB) in NE Iraq, and is considered a giant undiscovered Jurassic source rock [1]. Several studies have documented the main petrographic features of the Barsarin formation in order to characterize the paleoenvironment, but without any attention to diagenesis. The formation comprises laminated limestones and dolomitic limestones, in places cherty, with autoclastically brecciated beds admixed with shaly and marly materials [2]. The micro crystalline quartz (chert) associated with carbonate rocks have different silica sources, including a biogenic origin [3], silica-enriched seawater [4], and/or linked to silica input via a river source [5].

The origin and evolution of the diagenetic fluids that affected the Barsarin sediments are still largely unknown due to a lack of detailed studies. However, the fluids can be traced by geochemistry, and particularly by analyses of stable (carbon and oxygen) and radiogenic (strontium) isotopes. Oxygen isotope compositions can provide insights into the origin of mixed fluids or cross-formation water flows, and also to characterize paleo-temperatures and rainfall properties, such as the amount, seasonally, and moisture sources (e.g., [6–8]). Carbon-isotope compositions can provide information on effective rainfall and weathering processes [8]. The conventional $^{87}Sr/^{86}Sr$ isotope composition is classically used to infer the main sources of radiogenic strontium in sedimentary basins related to the pulses of mid-oceanic hydrothermal fluxes [9] or to infer overprinting by radiogenic dolomitizing fluids (e.g., [10]). It allows for the determination of continental riverine input [11]. However, the classical conventional technique to measure radiogenic isotopes is limited to the achievable spatial resolution of $^{87}Sr/^{86}Sr$ records.

To address the time specific gap in tracking the marine limestone and associated diagenetic fluids in the Barsarin formation, a new approach is developed here. It consists of coupling a high-resolution laser ablation (LA-) MC-ICP-MS analysis with fieldwork data, petrography, and geochemistry (oxygen-carbon stable isotopes). The remarkable high precision of laser ablation analysis is utilized to measure, for the first time in our sediments, the absolute radiogenic composition of the Jurassic carbonate rocks of the Barsarin formation, and evaluate the precise origin, involvement and evolution of different fluids during diagenesis that modified the original composition of the host limestones. This will also give information on the dynamic of the diagenetic fluids during the Zagros orogeny within the ZTFB.

2. Geological Setting

The Upper Jurassic high-folded zone, considered as a part of the Zagros fold-thrust belt, belongs to a NE-SW basin (Figure 1), inherited from the tectonic activity of the Arabian plate and the opening of the southern Neo-Tethys ocean [12]. This ocean was characterized by sea level fluctuations that affected evaporitic sabkhas in the basin during Jurassic-Cretaceous times. NE Iraq formed an euxinic basin separated from the Neo-Tethys by a rifted area where shallow water carbonates formed [12]. It is generally assumed that the Barsarin formation records an euxinic environment, subjected to a low subsidence that characterized the Neo-Tethys during this period.

The Barsarin formation is composed of laminated limestones and dolomitic limestones, and in places brecciated texture and crumbled and contorted beds have been also documented in the type locality and type section. The author also recognized a mixture of shaly and marly materials with melikaria structures, and the upper and lower contact boundaries of the Barsarin formation are identified by the Chia Gara formation and Naokelekan formation, respectively.

The Barsarin formation was previously described as an isolated lagoonal environment, mostly evaporitic, and based on stratigraphic position from fieldworks, it has a Kimmeridgian-Tithonian age [13]. The recent studied section of the Barsarin formation is located close to the Rawandus area where the series is considered as the reference section in northeastern Iraq (Figure 1).

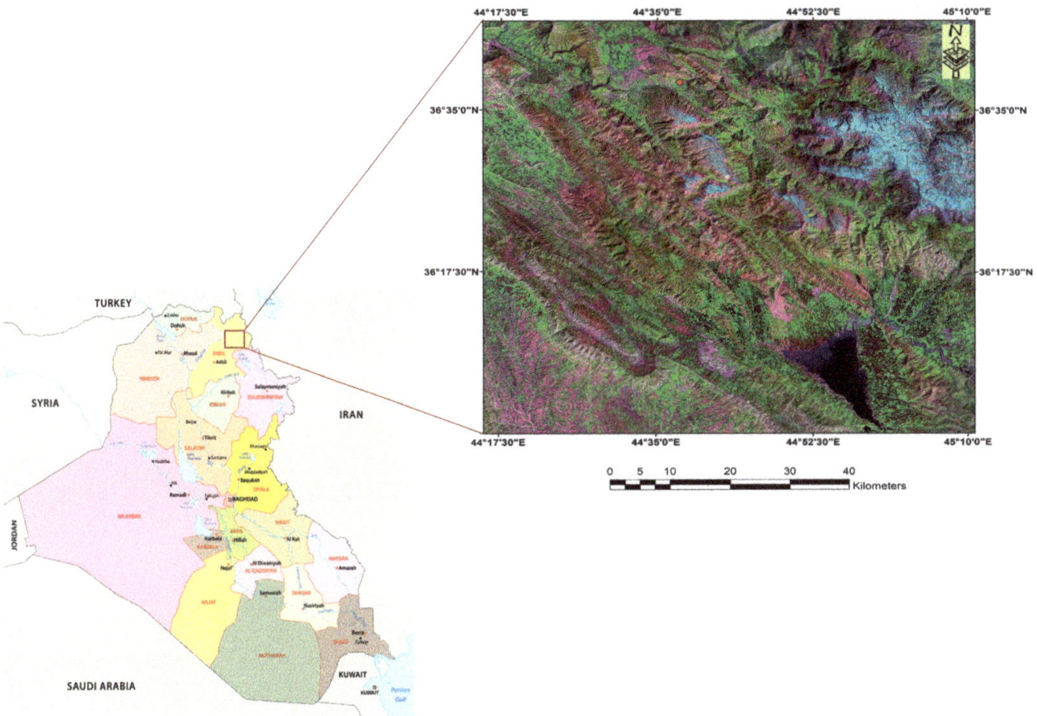

Figure 1. General map illustrates the location of the studied area "green circle".

3. Methods

Twenty-three samples were collected from the outcrop of the Barsarin formation. All the samples were prepared in laboratory for thin sections and studied under the optical microscope to distinguish the different carbonate phases, with a particular attention on the geometric cross-cut relations of fractures and veins. Scanning electron microscopy (SEM) was also used to image the surfaces to provide information about the morphology or texture of the surface. Selected samples of broken fragments (chips) mounted on aluminum stubs were studied from samples coated with gold depending on the purpose of work. SEM and EDX are also utilized in the recent study, backscattered image mode was used to give the different in contrast between minerals with different atomic number. The low atomic number samples give low emissions of backscattered electrons, while high atomic number samples give high emissions of these electrons. The backscattered electrons have higher energies than secondary electrons—usually from approximately 8×10^{-18} J (50 eV) up to the energy of the primary beam electrons.

The oxygen and carbon isotopic compositions of (24) samples were analyzed from powders, after selective microdrillings of the different recognized carbonate phases. However, the co-occurrence of fracture-filling calcite and dolomite, and also evaporites in the host limestones, made it difficult to avoid a mixing between these phases. As a consequence, the dolomite and calcite are considered as one group, and host limestone and evaporite as a second group.

Carbonate powders were reacted with 100% phosphoric acid at 70 °C using a Gasbench II connected to a Thermo Fisher Delta V Plus mass spectrometer. All values are reported in per mil relative to V-PDB (Table 1). Reproducibility and accuracy were monitored by replicate analysis of laboratory standards calibrated to international standards NBS19, NBS18 and LSVEC. Laboratory standards were calibrated by assigning $\delta^{13}C_{VPDB}$ values of

+1.95‰ to NBS19 and −46.6‰ to LSVEC and by assigning $\delta^{18}O_{VPDB}$ values of −2.20‰ to NBS19 and −23.2‰ to NBS18. The analyses (23) were performed in the University of Erlangen.(Germany, M. Joachimski).

Table 1. $\delta^{13}C$ (% VPDB), $\delta^{18}O$ (% VPDB) values of selected samples from Barsarin formation (n = 24).

Sample No.	$\delta^{13}C_{VPDB}$	$\delta^{18}O_{VPDB}$
B.4	−9.3	−0.8
B.15	−7.5	−2.9
B.19	−5.6	−5.9
B.5	−9.6	−3.5
B.3	−8.1	−1.3
B.11	−8.2	−3.8
B.11	−8.4	−3.9
B.12	−7.0	−1.4
B.8	−10.7	−4.3
B.0	−8.7	−0.3
B.0	−5.1	−12.5
B.6	−4.9	−12.9
B.6	−11.7	−5.1
B.22	−8.2	−1.2
B.B	−8.6	−9.6
B.9	−2.5	−11.3
B.20	−6.3	−1.7
B.21	−5.2	−2.3
B.14	−7.5	−0.8
B.14	−6.2	−6.4
B-Contact	−10.1	−4.1
B.16	−7.1	−2.3
B.1	1.5	−7.5
B.10	−11.0	−4.9

Strontium isotope measurements on carbonate samples were performed by Laser Ablation ICP-MS at Goethe University-Frankfurt, using a Thermo-Finnigan Neptune multicollector (MC)-ICP-MS system attached to a Resolution 193 nm ArF Excimer laser ablation system (ComPexPro 102F, Coherent), equipped with an S-155 two-volume (Laurin Technic, Australia) ablation cell. Square laser spots with diameters or edge lengths of around 235 μm were drilled with an 8-Hz repetition rate, and energy density of about 6–7 J cm^{-2}, during 45 s of data acquisition.

The collector set-up included (1) ^{83}Kr as a monitor to verify the successful correction for the isobaric interferences of ^{84}Kr on ^{84}Sr and of ^{86}Kr on ^{86}Sr by subtraction the gas blank measured before sample ablation; (2) mass 83.5 to gauge the production rate of doubly charged ^{167}Er and calculate the production of doubly charged ^{164}Er, ^{166}Er, ^{168}Er and ^{170}Er, which interfere with ^{83}Kr and ^{84}Sr, ^{85}Rb and ^{86}Sr, respectively; (3) mass 86.5 to gauge the production rate of doubly charged ^{173}Yb and calculate the production of doubly charged ^{168}Yb, ^{172}Yb, ^{174}Yb and ^{176}Yb, which interfere with ^{84}Sr, ^{86}Sr, ^{87}Sr and ^{88}Sr; (4) ^{86}Sr and ^{88}Sr to use $^{88}Sr/^{86}Sr$ for mass bias correction; (5) ^{84}Sr to check the accuracy of the mass bias and interference (see above) correction using $^{87}Sr/^{86}Sr$; (6) ^{85}Rb to correct for the isobaric interference of ^{87}Rb on ^{87}Sr and to get the $^{87}Rb/^{86}Sr$; and (7) ^{87}Sr to use in the radiogenic isotope ratio $^{87}Sr/^{86}Sr$.

At the beginning of the analytical session, a soda-lime glass SRM-NIST 610 was measured to three times for empirical determination of 87Rb/85Rb mass bias. The procedure yielded, after interference correction on ^{86}Sr, ^{88}Sr, and ^{85}Rb from doubly charged Yb and Er, the $^{87}Rb/^{86}Sr$ ratio needed for accurate correction of the isobaric interference of ^{87}Rb on ^{87}Sr. An isotopically homogeneous in-house plagioclase standard (MIR-1) was measured throughout the analytical session to monitor accuracy and apply corrections to the measured unknowns, if necessary. This plagioclase is a megacryst from a lava of the

Dutsin Miringa Hill volcano (Northern Cameroon Line; [14]) that has been independently characterized for $^{87}Sr/^{86}Sr$.

The reference materials BHVO-1 and BCR-2G were measured along with the in-house standard to monitor accuracy, in particular of the correction of the isobaric interference of ^{87}Rb, and also the reproducibility of the results. The corrections for the presence of Rb were minor and the results for the unknowns, reported in Table 2, are considered to be accurate. Sr concentrations compared with MIR (3500 ppm), applying the same corrections as for MIR. All analyses together are given in Table 2.

Table 2. Strontium concentration values and $^{87}Sr/^{86}Sr$ ratios by laser spots analyses using ICP-MS (B9, B14, n = 35).

Sample No.	Line No.	Sr ppm	$^{87}Sr/^{86}Sr$
B9	109.00	148	0.70767
B9	110.00	160	0.70772
B9	111.00	191	0.70767
B9	112.00	690	0.70750
B9	113.00	496	0.70744
B9	114.00	154	0.70768
B9	115.00	594	0.70743
B9	116.00	578	0.70745
B9	117.00	1029	0.70746
B9	118.00	528	0.70739
B9	119.00	208.51	0.70789
B9	120.00	419	0.70730
B9	121.00	642	0.70747
B9	122.00	144	0.70744
B9	123.00	127	0.70801
B9	124.00	44	0.70767
B9	125.00	243	0.70763
B9	126.00	288	0.70743
B9	127.00	193	0.70725
B9	128.00	202	0.70721
B9	129.00	690	0.70755
B9	130.00	359	0.70747
B14	131.00	455	0.70755
B14	132.00	434	0.70750
B14	133.00	508	0.70749
B14	134.00	471	0.70746
B14	135.00	362	0.70747
B14	136.00	163	0.70731
B14	137.00	163	0.70731
B14	138.00	266	0.70735
B14	139.00	75	0.72859
B14	140.00	224	0.71131
B14	141.00	235	0.70761
B14	142.00	234	0.72629
B14	143.00	71	0.70721

4. Results

4.1. Field Observation

The base of the Barsarin formation is characterized by decimetre-thick beds (<30 cm) of stromatolites, evaporites (Figure 2B,C) and a prominent cherty level, alternating with well-bedded dolomitic limestones and shaly limestones (Figure 2A). The top of the formation shows massive limestones and dolomitic limestones (Figure 3). The thickness of the formation is 8 m in the studied Barsarin section type.

Figure 2. (**A**) Photograph showing the exposed section of the well-bedded limestone deposited during Upper Jurassic, Barsarin formation. (**B**) Photomicrograph illustrates the micro-laminated features of Bositra-like pelecypod shells (see [15]), the micro-laminated features cross-cut by vein of evaporites; (**C**) collapsed mudstone microfacies, giving auto-brecciated sediment. The pore spaces of the breccia are completely occluded by evaporates, with no traces of late diagenetic cements.

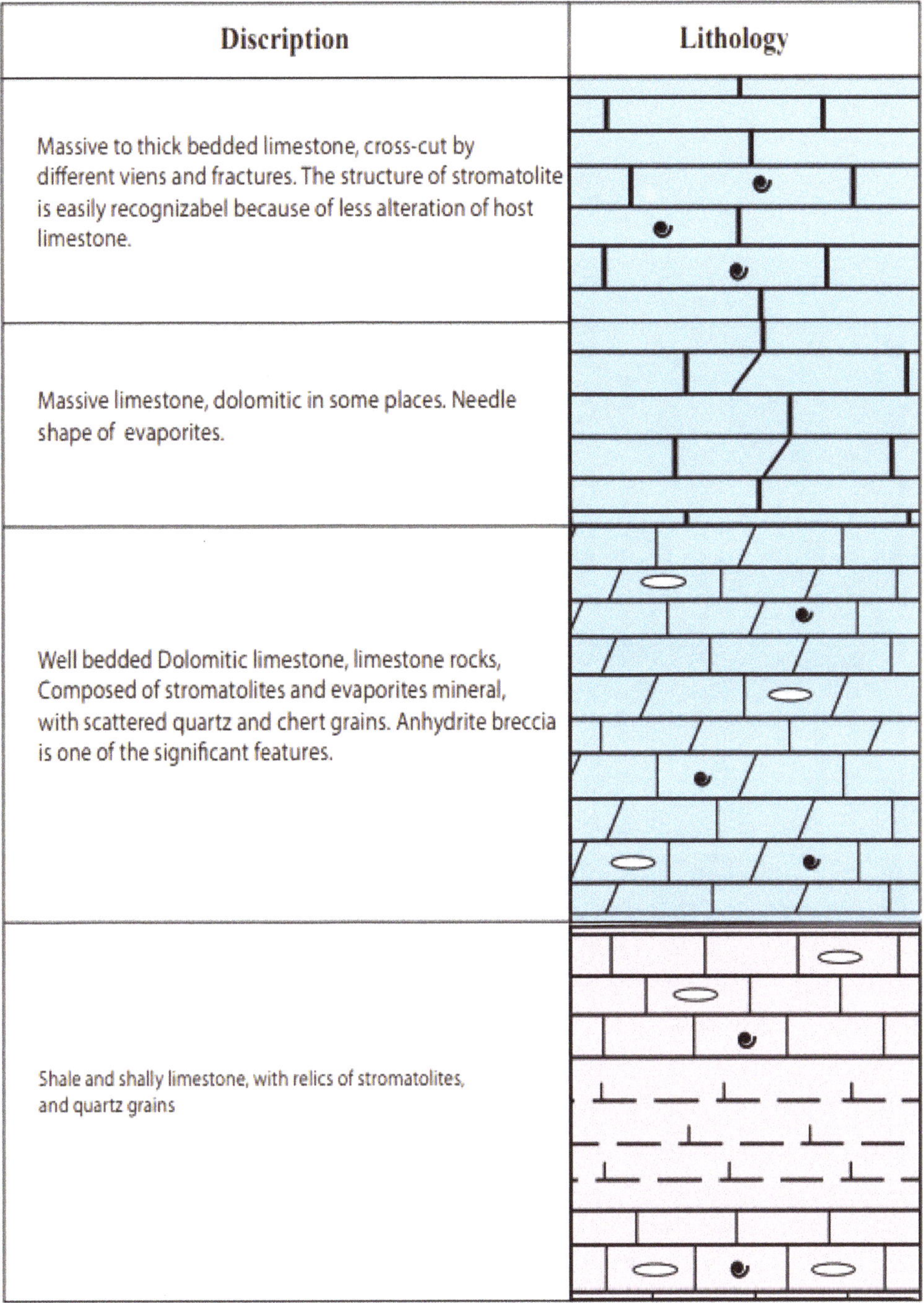

Figure 3. Master log showing the general lithology of the Barsarin Formation (Not to scale).

The fracture- and void-filling dolomite and calcite are distributed throughout the Barsarin profile, in addition to several vertical calcite filling joints and few horizontal ones in the upper part.

4.2. Petrography

The host carbonates show three types of microfacies: (i) mudstones, (ii) radiolarian wackestones and (iii) stromatolite boundstones (Figures 3 and 4). Despite these facies having undergone dolomitization, they are still recognizable, particularly the fine Bositra-like pelecypod shells and stromatolitic fabrics in the lower and parts of the section (Figure 2B).

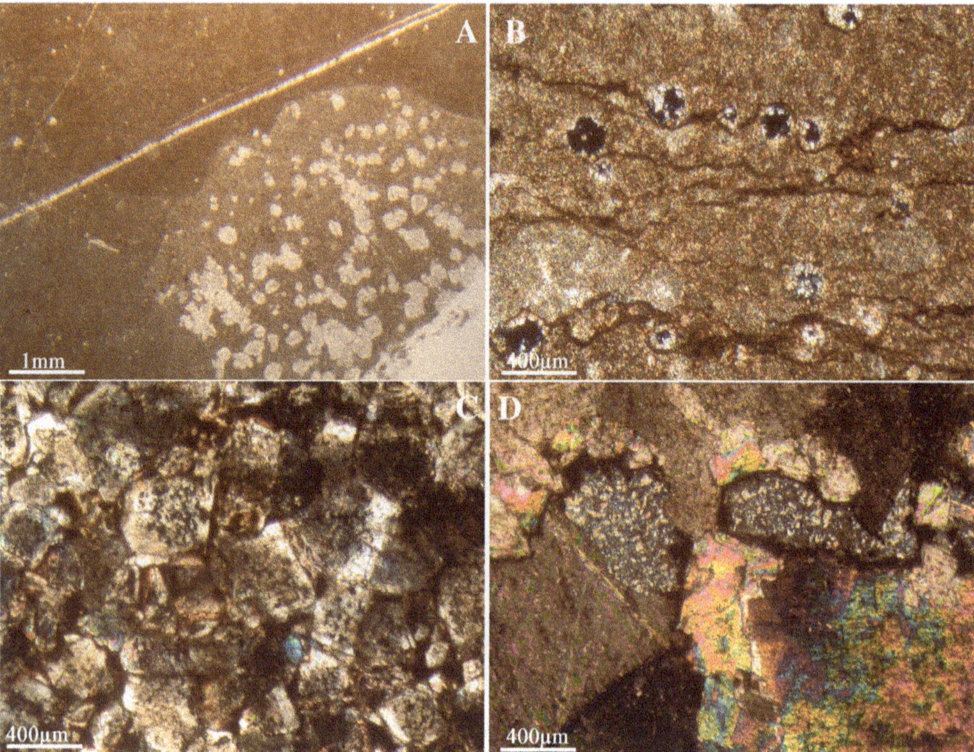

Figure 4. Photomicrographs illustrating characteristics of silicate mineral content. (**A**) Litho-clast "right most part of the photo" embedded by mega-quartz grains. (**B**) Micro-organisms of spherical shapes of Radiolaria, the chalcedony micro-crystalline quartz partially filled the pore spaces of Radiolaria. (**C**) Mega-quartz grains in places replaced by dolomite minerals, see the strong relief and inclusions of quartz in the core of some grains. (**D**) Late diagenesis dolomite cements "D_b" replaced the microcrystalline quartz.

Abundant needles/lath crystals from evaporite minerals are found within the mudstone microfacies, which shows an in-situ brecciated fabric with irregular cracks and veins (Figure 2C). Fracture-filling evaporites are volumetrically less important than the needle shaped evaporites (Figure 2A). The in-situ collapse of the pristine facies leading to a breccia is related to the dissolution of the evaporitic minerals (Figure 2C). Sometimes, these breccias are still filled with evaporite minerals.

Dolomite appears as a cement in two forms. D_m represents the initial precipitate, and it is characterized by dark coloured crystals of variable sizes and anhedral-euhedral shapes (Figure 5B,C). D_m shows, in places, a typical rhombohedral shape, with micritic-rich core and transparent cortex (Figure 5B). Larger transparent euhedral crystals are also observed. The second cement occurs in the fractures and vugs and consists of a dirty coarse euhedral dolomite (D_b) with curved or non-planar surface. It can also be euhedral with a characteristic zonation and shows a sweeping extinction under crossed nicols (Figure 5C,D). D_b only occurs in fractures and open spaces. Fracture-filling calcite crystals are always associated with the second dolomite (D_b); it displays variable sizes, is transparent and consists of blocky, twinned crystals. Geometric relationships under petrographic analysis show that the blocky calcites were replaced by coarse dolomite (D_b). In addition, the detrital grains, mainly quartz, are observed within the host carbonate. They show the traces of dissolution and erosion, and this could be due to the distance of grain transportation (Figure 6A,D).

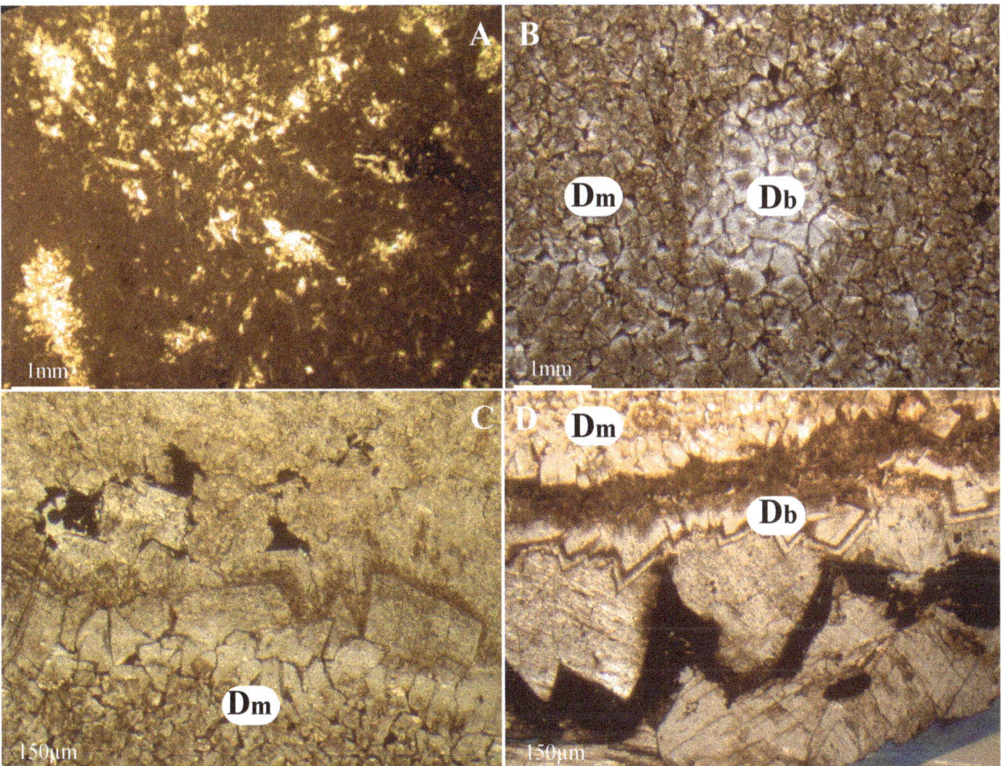

Figure 5. Photomicrographs illustrate: (**A**) The needle shape of evaporites within the mudstone microfacies that filled the pore spaces, (**B**) early dolomitizing cement (D_m) were mostly preserved the previous pristine facies "dark brown colour", while in the central part of the photo the rhombohedral shape of dolomite which precipitated later. The lower part (**C**) and upper part (**D**) of the photos show the predating cement formation "D_m" and the coarse grains of dolomite "D_b" from the late dolomitizing fluids.

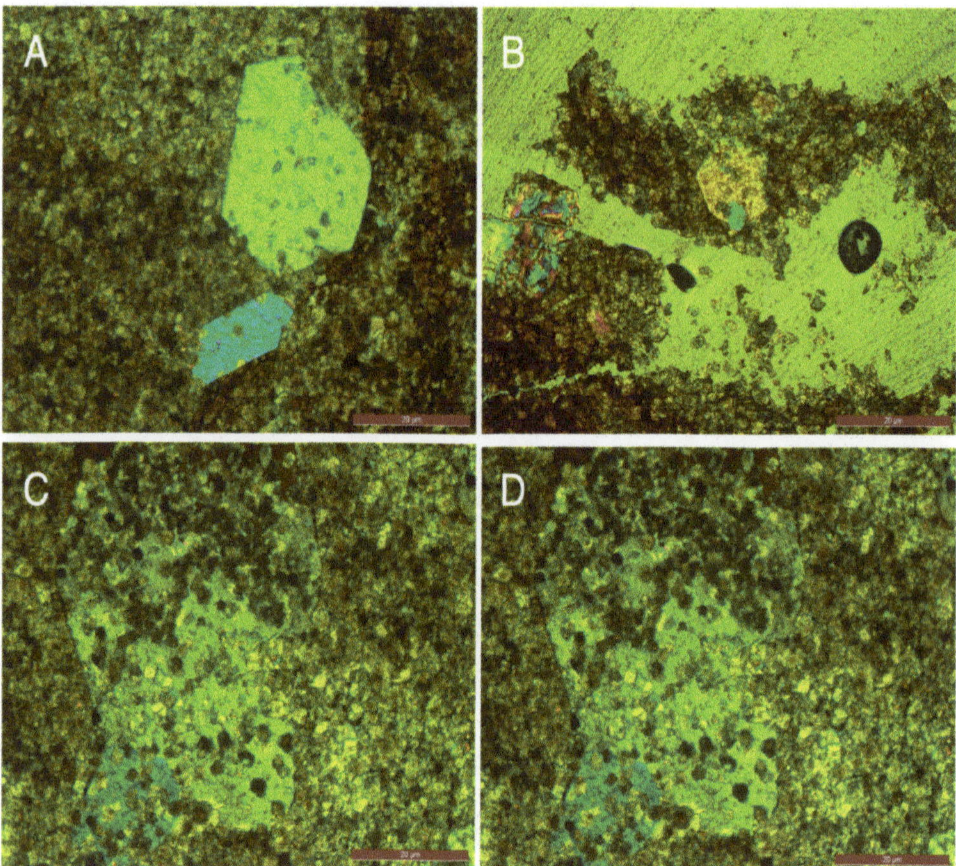

Figure 6. (**A–D**) Photomicrographs illustrate the weathered clasts of quartz from outside of the Barsarin basin, close up the eroded outline of polygonal grains (**B**) and other detrital grains in (**D**,**C**) that embedded in host carbonate rocks.

SEM analyses reveal that the host limestone, dolomite and calcite cements are commonly associated with quartz grains, more specifically at the bottom and upper parts of the formation (Figure 7). The grain size of quartz ranges from micro-quartz to mega-quartz, and in places silica appears in its chalcedony form.

4.3. Oxygen, Carbon and In Situ Strontium Isotopes

The O–C isotope compositions of the pristine facies, calcite, dolomite and evaporite are listed in Table 1. The 23 samples from fracture-filling dolomite and calcite cements, matrix dolomite (D_m) and pristine facies of the formation have oxygen–carbon isotopic values populated into two groups. Group I is isotopically heavier in composition than group II (Figure 8). The host limestone and evaporite oxygen–carbon values fall into group I, while the calcite and dolomite cement oxygen-carbon values mostly fall into group II.

Most of the carbon and oxygen marine isotopic compositions of the Barsarin facies are significantly lower than those of seawater limestone during Late Jurassic time reported by Brand and Anderson [16,17]. Isotope compositions of Late Jurassic belemnites led these authors to define the isotopic composition of seawater during this time, with $\delta^{18}O_{VPDB}$ values ranging from 0.04‰ to −0.99‰ and $\delta^{13}C_{VPDB}$ values from 0.97‰ to 1.51‰. In group I of the Barsarin samples, $\delta^{18}O_{VPDB}$ values are between −5.09‰ to −0.34‰, and

$\delta^{13}C_{VPDB}$ between −5.23‰ to −11.66‰. In group II, $\delta^{18}O_{VPDB}$ values are −12.87‰ to −5.90‰, and $\delta^{13}C_{VPDB}$ values are −8.55‰ to +1.47‰. However, a few host limestones that are free of evaporites have lighter oxygen isotopic values (up to −4.34‰) than those containing evaporites (up to −2.33‰, see the in Table 1).

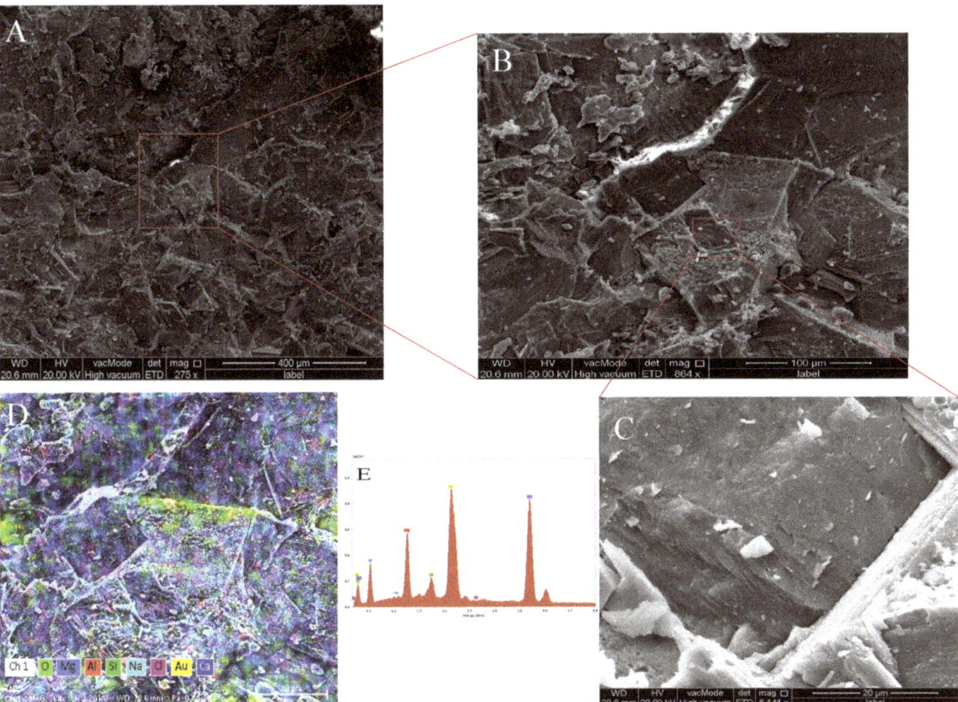

Figure 7. SEM/EDX images of the evaporite carbonate rock sample obtained from freshly gold coated surface, the images illustrate: (**A**) the carbonate grains cemented by evaporitic minerals, the circle marks the aggregate grains of gypsum crystals, the rectangle marks two phases of minerals and this part, enlarged in (**B**,**C**), where the central part, composed of euhedral, eroded the surfaces of calcite crystals where cemented by evaporites. (**D**,**E**) Mineralogical mapping of the same sample location (**C**); (**D**,**E**) the majority of sodium element was distributed around the euhedral crystals of calcite.

Two samples (B9, B14, Table 2) have been analysed by laser ablation to obtain an absolute high-resolution radiogenic $^{87}Sr/^{86}Sr$ ratios on fracture-filling cements and the carbonate matrix. In-situ measurement of $^{87}Sr/^{86}Sr$ signatures in calcite and dolomite are provided by laser ablation to obtain the absolute radiogenic composition of many measurements on a micrometre-sized scale (MSS) within a single crystal (Figures 9–11). Laser spots analyses by ICP-MS measured several precise points on each crystal in two samples (B9, B14, n = 35). The plotted$^{87}Sr/^{86}Sr$ ratios display two positive excursions (Figures 10 and 11). The first excursion is characterized by the highest radiogenic signatures, ranging from 0.70789 to 0.72859. Dolomite and calcite crystals fall into the second positive excursion, with strontium isotope ratios varying from 0.70721to 0.70772. They are less radiogenic than the host limestone values. Most of the laser spot measurements from the host limestone show a significant Sr radiogenic ratio, and mostly fall into the first positive excursions. The absolute strontium isotope data of the host limestone is radiogenically very high compared to the data from literature [9], and do not fit with the late Jurassic seawater strontium isotope curve, which varied between 0.70671 and 0.70713.

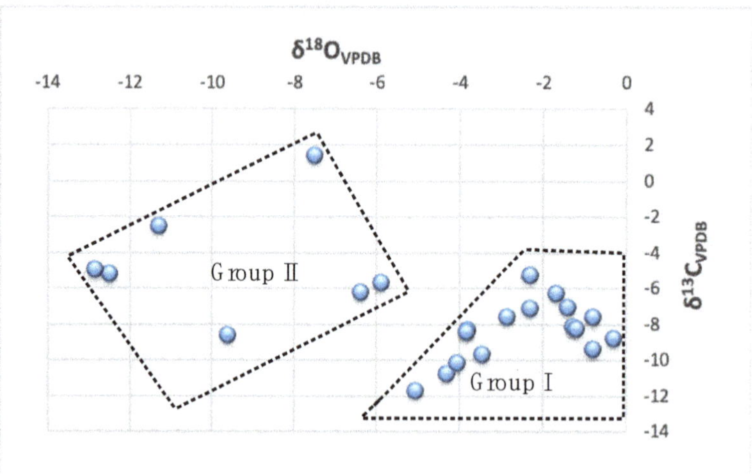

Figure 8. Oxygen and carbon isotope compositions in the Barsarin formation (n = 24). The values are populated on two groups: the first, group I (evaporitic mudstones and wackestones), has heavier values than the second, group II (calcite and dolomite cements).

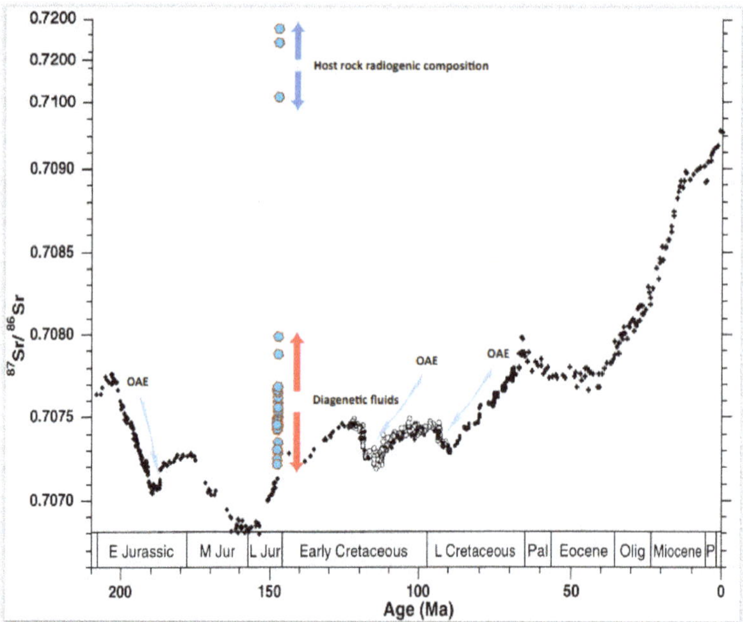

Figure 9. The seawater Sr-isotope curve shows minima in the Late Triassic, Pliensbachian-Toarcian, Callovian-Oxfordian, Aptian-Albian, and Cenomanian-Santonian. The recent study data "blue plots" from our study are shown for comparison. The recent data "this study" populated around two positive excursions: the first host carbonate positive excursion, and the second diagenetic carbonate positive excursion. The host carbonate value is highly radiogenic, higher than the whole seawater Sr isotope curve, while the value of carbonate diagenesis samples is still high and less radiogenic than host carbonate samples (modified after Jones and Jenkyns [9]). OAE = Oceanic Anoxic Event.

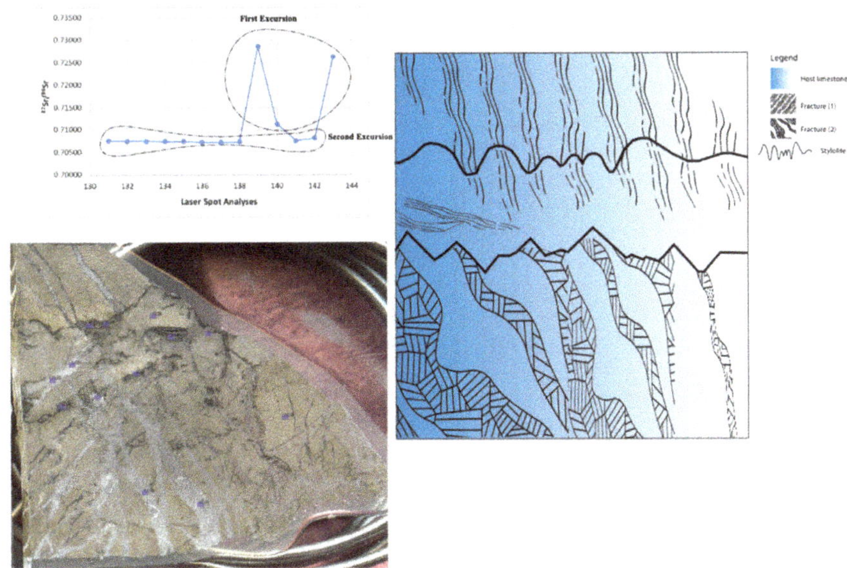

Figure 10. The absolute radiogenic strontium isotope values v. the location of the spot analysis "sample No. B14". The location of analysis is represented by cubic blue colour where the ablated spot size used is 213 μm and depth of crater is ~20 μm.

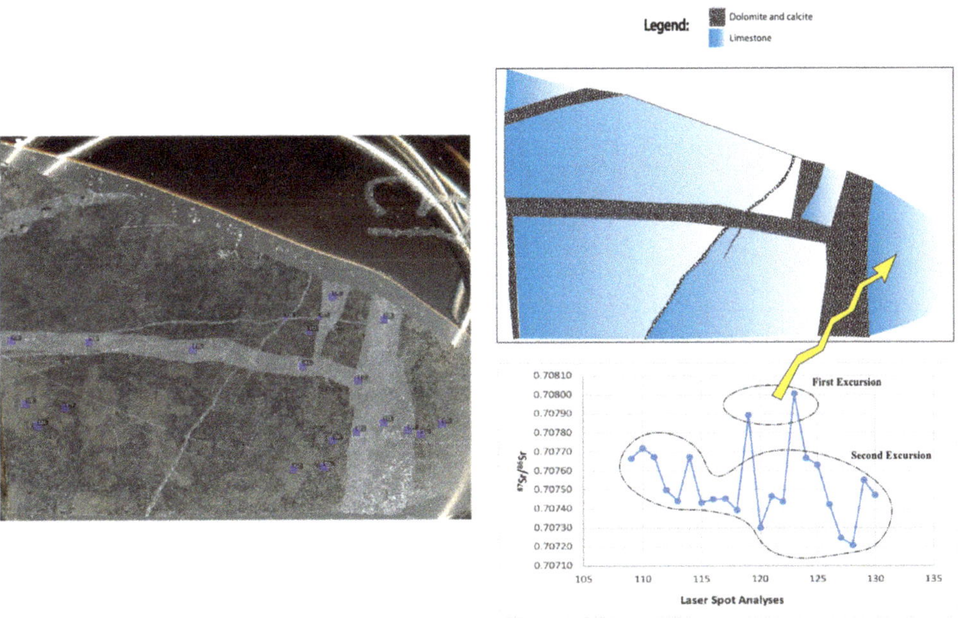

Figure 11. The absolute radiogenic strontium isotope values v. the location of the spot analysis "Sample No. B9". The location of analysis is represented by cubic blue color where the ablated spot size used is 213 μm and depth of crater is ~20 μm.

5. Interpretation and Discussion of Conventional and Non-Conventional Isotopic Signature Trends

5.1. $\delta^{13}C_{VPDB}$-$\delta^{18}O_{VPDB}$ Isotopes

The estimation of isotope values of marine and diagenetic sediments is useful in providing a better understanding of seawater compositional changes during deposition or post-deposition due to any chemical or biological alteration. The oxygen isotope reconstruction in Mesozoic marine sedimentary rocks shows a general decrease of $\delta^{18}O_{VPDB}$ values of the global seawater throughout the Late Jurassic [16–18]. Brand and Anderson proposed a past seawater $\delta^{18}O_{VPDB}$ from around +2.5‰ to −3‰ during Late Jurassic. Furthermore, the general trends of a warming through the Late Jurassic yields a paleo-temperature value in the range of 20–30 °C [19]. $\delta^{18}O_{VPDB}$ values between −5.09‰ and −0.34‰ of our host limestone (Group I; Table 1) do not fit within the previously documented oxygen isotope values of marine carbonate during Late Jurassic and is probably related to paleo temperature conditions.

Paragenetically, our host carbonate rocks from the Upper Jurassic have significant negative values of oxygen and carbon isotopic compositions, as low as −5.09‰ and −11.66‰, respectively. The host carbonates predated evaporite formation (see Figure 7 and Group I from Figure 8). Host carbonate samples that contain relicts of evaporites have $\delta^{13}C_{VPDB}$ $\delta^{18}O_{VPDB}$ values (from 0.34‰ to −2.33‰ and −6.25‰ to −9.30‰, respectively) lower than those of the previous reported marine domain for the same period [20], but higher than those of the pure host carbonates in our study. This trend of wide variation in oxygen and carbon isotope compositions from the host carbonates might be due to fluctuations of sea level, possibly related to climate change, at a local, regional scale (e.g., [21]), if not at a global scale [9], and/or could be linked to a secondary imprint on marine rock composition.

Floating angular clasts of host carbonates are cemented by evaporite crystals, suggesting the evaporative episode (mainly gypsum and halite) post-dated the host carbonates (Figures 2C and 7). This led to the formation of collapse breccias from the host carbonates, probably related to sea-level fall and climate change, as measured previously by the ratio of precipitation to evaporation (e.g., [22]). Hence, the large depletion in oxygen and carbon values, together with systematic occurrence of evaporates, suggest that dry and warm climate conditions drove the primary oxygen signal to negative values, where oxygen isotopic composition of seawater in the study basin increased during precipitation (e.g., [23]).

Evaporite minerals reflect the hypersaline evolution of the basin that led to heavier oxygen isotopic values than those of marine limestone (e.g., [24]) [25] in contrast to the carbon isotopic composition [26]. Nevertheless, our ^{18}O depleted values in Group I are lighter than those of the oxygen isotopic composition of the marine curve. $\delta^{18}O_{VPDB}$ depleted values in host carbonates (Group 1) reveal that the former evaporative episode was affected by climatic changes during Late Jurassic, possibly through river input during a humid season [26]. Therefore, the depleted oxygen values from our study, combined with published data for the Late Jurassic-Early Cretaceous interval, suggest that global and gradual warmer palaeotemperatures occurred during this interval when the Barsarin formation was deposited. A similar scenario has previously been reported by Lécuyer and Allemand [27] and Vickers [28], from clumped isotopes and conventional oxygen isotope data, to interpretate the climatic change throughout Jurassic Cretaceous times. Furthermore, the covariance between oxygen and carbon isotopic compositions in marine carbonates (Figure 8) is probably linked to period boundaries with climate change [29,30].

The O–C isotopic values of second group (Group II) are lighter in their oxygen and heavier in their carbon isotope composition than those in Group I, although some of the carbon values overlap the marine signature. Group II has distinct and wide variations of oxygen–carbon isotopic compositions, compared to Group I. Group II is influenced by late diagenetic fluids that had a role in precipitation of calcite and dolomite cements (Figure 5). The $\delta^{18}O_{VPDB}$ values of diagenetic cements exhibit more depleted values than those not influenced by diagenetic fluid alteration (Figure 8). This could be due to the long-term interaction between the host rock and the diagenetic fluids, promoted by fluids heated up to 80 °C or more (e.g., [31]). Therefore, the considerable and significant variation of $\delta^{18}O$–$\delta^{13}C$ from Group II could be linked to the evolution of fluids (e.g., [31]), with slow precipitation rates under subsurface conditions (e.g., [32]), or rapid precipitation of dolomite and calcite due to abrupt change in pressure and temperature [8,33]. The extremely negative $\delta^{18}O$ values in the coarse-sized dolomite and calcite crystals (Figure 8) highlight the interaction of the host limestones with diagenetic fluids (e.g., [8]) as indicated in Group II. Since the oxygen isotope values are extremely negative in deep diagenetic conditions, where the mechanism for dolomite and calcite precipitation is slow, hot circulation is the most probable source for samples belonging to Group II. Nevertheless, Group II is populated in two $\delta^{18}O$ assemblages. One follows the J trend of meteoric water with extreme depleted $\delta^{13}C$ values, and the second shows lighter oxygen isotope compositions with slight changes in $\delta^{13}C$ values. These are quite consistent with the mixing of hot and meteoric fluids.

The carbon isotope values in Group II also exhibit a wide variability in their composition, although some values still possess host rock signatures (Group I: Figure 8). This could be related to the buffered system during deposition of diagenetic cements (e.g., dolomite and calcite) where the diagenetic fluid mixed with rock-derived carbon of the previous carbonates (Group I).

Considering the quartz grains in the host carbonate, the lower and upper parts of Barsarin formation are richer in silica than those of the middle part. Since the oxygen isotope composition of carbonates rich in silica is lighter than seawater, the chemistry of water facies rich in silica is not related to a marine source. Therefore, the origin of the quartz grains is not related to diagenetic evolution, as previously reported in a similar case by Bustillo [34], despite the outer rims of the chert or quartz having been often replaced by dolomite/calcite in a late diagenetic event (Figure 4D). This case has been reported also by Bustillo [35].

5.2. Sources of Variation of Absolute Strontium Isotopes from the Late Jurassic and Onwards, Utilizing Non-Conventional Laser Ablation ICP-MS

Seawater $^{87}Sr/^{86}Sr$ ratios varied homogeneously through geological time, thus making this isotope proxy an important chemostratigraphic tool [36,37]. The evolution of the seawater $^{87}Sr/^{86}Sr$ ratio curve during the Jurassic-Cretaceous is widely driven by variation in continental weathering rates (riverine) and predominant oceanic fluxes of non-radiogenic strontium sporadically during the Jurassic [5]. Silica hosted carbonates are another source that impacted the seawater radiogenic composition curve [38]. These sources of radiogenic strontium isotopes had a different effect on the seawater strontium isotope record (Figure 8, [5,9]). Our data provides, for the first time ever, a high-resolution strontium isotope composition from pristine facies and associated diagenetic carbonates using non-conventional direct laser ablation (Figures 10 and 11) in Upper Jurassic carbonate source rock.

The diagenetic imprint has to be precisely considered since the diagenetic fluids are impacted by the uptake or loss of Sr radiogenic isotope and Sr concentration. Therefore, the petrographical characteristics and paragenetic sequence for the entire section is critical before interpreting the strontium isotopic analysis. In the Barsarin formation early silicification and evaporite minerals formation is possibly syn-depositional [39] or occurred during a very early diagenetic event (e.g., [35]). The collapse and brecciation of the mudstone facies suggest that the host sediments and evaporite minerals were coincident with the depositional time of Barsarin formation. Silica infilling the intragranular porosity is also observed within dolomite and calcite lattice and suggests a late diagenesis process. The co-occurrence of silica and diagenetic carbonates bring the idea of silica recycling and/or continuous flux "input" of silica during and after the depositional setting.

The $^{87}Sr/^{86}Sr$ ratios are populated into two positive excursions. The first excursion gives a radiogenic isotope value of up to 0.72859 in the host carbonate rocks; these values are three times higher than those reported during Late Jurassic times [9]. The second excursion observed in the calcite and dolomite cements is less radiogenic than host carbonates. The core of carbonate cement shows impurities due to growth of quartz grains in their crystal framework (Figure 4C). These two positive excursions of absolute strontium radiogenic data could be explained in the following two ways:

(i) The silicate sediments that precipitated from seawater were converted into fibrous quartz and/or micro- (i.e., the biogenic silica-rich sediments) and mega-quartz via dissolution–reprecipitation processes [40]. The first excursion in pristine facies, that often contains silicate minerals in the studied samples, has lower Sr concentration and higher radiogenic strontium isotope values (high silicate minerals). The silicates were observed through the whole thickness of the carbonate facies, and as well as solid inclusions within the core of the calcite and dolomite crystal lattices. These inclusions could be the reason for the enrichment of the strontium ratio in the carbonate cements. Therefore, silica cycling via dissolution–reprecipitation processes could be one possible scenario for the observed radiogenic strontium, mainly for the first positive excursion (Figures 10 and 11). The silicate sediments in host carbonate show no corrosion or disturbed grains due to dissolution of the siliceous sediments or replacement, but in places the groundmasses are embedded with silicate grains totally different from the host carbonate facies, as highlighted frequently by their different colours from those of the facies (Figure 4). This indicates that some of the silicate sediments likely derive from outside of the Barsarin depositional basin, i.e., they are siliceous sediments continental detritus (Scenario II).

(ii) The second positive excursion relates to the diagenetic cement, where the strontium isotope ratio and strontium contents formed two assemblage groups: the highest strontium ratios with lowest strontium concentration (first group) and the lowest strontium ratios with highest strontium contents (second group). In addition, the weathered and eroded particles embedded in host carbonate (Figure 6A–D) reveal that this phase it could be derived from the mixing of detrital silicate minerals transported by rivers from the catchment "low Sr content" with high diagenetic Sr content "hot fluids".

The strontium isotopic composition of seawater is predominately controlled by inputs at mid-ocean ridges, through hydrothermal exchange ($^{87}Sr/^{86}Sr$ = ~0.703), and from rivers, through continental erosion ($^{87}Sr/^{86}Sr$ = 0.705 to > 0.800; mean ~0.712; Palmer and Edmond, 1989). Variations in seawater ($^{87}Sr/^{86}Sr$ reflect changes in the concentration of Sr and the $^{87}Sr/^{86}Sr$ of dominant sources, [18]), with the hydrothermal source being less radiogenic (~0.703), generally account for less of the total Sr input in the ocean system [41]. The concentration and isotopic value of the riverine flux is highly variable due to heterogeneities in the isotopic composition of continental crust, the concentration of Sr in source rocks and weathering rates.

The increased input of hydrothermal strontium at oceanic ridges had a major effect on seawater strontium isotope composition during the Jurassic [11]. Hydrothermal sources decreased seawater $^{87}Sr/^{86}Sr$ ratios sporadically during the Bajocian–Callovian period, while, during mid-Bajocian, elevated submarine volcanism increased the $^{87}Sr/^{86}Sr$ ratios (cf. [36]; Figure 9). The enhanced hydrothermal strontium input culminated with the global sea-level rise at the Middle-Late Cretaceous [42]. However, more published data showing a degree of scatter confirm the uniform rise of seawater strontium isotopic curve from Late Jurassic–Early Cretaceous times [23]. During the Late Jurassic, major events, like early Mesozoic rifting and Jurassic subduction, have been suggested to occur during the Zagros Orogeny [43]. Stratigraphic and geochronological keys suggest that a large volume of detrital continental crust contaminated by mantle-derived magmas impacted Late Jurassic times [44]. These authors concluded that the strontium enrichment was either derived from an enriched mantle source or acquired by crustal assimilation. Although, by the Late Jurassic the highest $^{87}Sr/^{86}Sr$ ratio reported by Lechmann [44] was as large as 0.70697, which is very low compared to our study (values reaching 0.72859).

The widespread values along one host limestone sample's "first excursion" has a higher radiogenic isotope ratio, of as much as 0.72859 (Figure 10), although the ratio in the seawater reference curve did not exceed 0.70720 during Late Jurassic times (Figure 9). Davies and Smith [45] documented that the strontium isotope ratio increased relative to pristine values by releasing strontium through the interaction of hydrothermal "hot" fluids with siliciclastic basement, or by release from continental siliciclastics [9]. The latter source is considered as radiogenic when evolved from riverine input [46]. Linking the continental siliciclastic source with our petrographical observations, the host carbonate rocks of the Barsarin formation show that silicification came from radiolarian cherts (Figure 4B). Yet, Radiolaria are not sufficiently abundant to be a sink for strontium isotopes. The early diagenetic siliceous cement is related to the source of silica from Radiolaria [47], and the primary intraparticle porosity was filled by chalcedony chert, as occasionally observed in our study area. Different-sized lithoclasts are embedded in the host carbonate rocks and contain variable sized-silicate grains (Figure 4A). These embedded silicate grains exclude that the radiogenic strontium was provided from the radiolarian cherts. Hence, at least the first radiogenic excursion was controlled by transport-limited nature of the weathering reaction, the same case was reported by Palmer and Edmond [41].

In SE France, hydrothermal dolomite, hosted in Upper-Jurassic sediments, were discovered with similar radiogenic values (ratios reaching 0.71120, Salih et al., 2020b). In this case the dolomite was related to radiogenic sources derived from riverine input. Furthermore, uplift in the latest Jurassic initiated detrital inputs related to the erosion along the Zagros Mountains [43]. Thus, the first positive excursion of an absolute strontium ratio would indicate an increase in weathering rates due to flux of radiogenic strontium.

The second positive excursion from the diagenetic cement samples evolved from diagenetic fluid sources. The dolomite and calcite cements preserved the inclusions and traces of silicate materials (Figure 4C). The remanent of solid inclusions inside the dolomite and calcite thereby limits the strontium isotope ratio. The remanent of silicate minerals either originated from host carbonates during replacement process or from interaction of hot fluids with radiogenic basement [5]. The origin of dolomite from hot fluids would preferentially record the interaction of hot fluids interaction with a radiogenic basement (cf., [5]), or the radiogenic source together with negative $\delta^{13}C$–$\delta^{18}O$ linked to seepage of low-temperature meteoric water in subsurface carbonate rocks [48].

5.3. Origin of $\delta^{13}C_{VPDB}$ $\delta^{18}O_{VPDB}$-Light and $^{87}Sr/^{86}Sr$-Rich Sediments

The enrichment and depletion of strontium and oxygen–carbon isotopes in carbonates is mainly controlled by sea-level fluctuation, the erosion of continental sediment and the recycling of ocean water through mid-oceanic ridges [49,50]. The absolute $^{87}Sr/^{86}Sr$ ratios of the host rocks and diagenetic cements in Barsarin formation show two positive radiogenic excursions, which are higher than those reported in the Upper-Jurassic marine carbonates [9]. The two positive excursions that have a linear extrapolation with Sr concentration, suggesting that $^{87}Sr/^{86}Sr$ ratios could be released from meteoric ($^{87}Sr/^{86}Sr$ ratios of up to 0.72859 and Sr concentration of 75 ppm) in the host carbonate rocks (Figure 12).

The deviation of our Sr isotopes values from the Sr isotope-seawater curve could be linked to sea-level fall during Late Jurassic, with a relatively warmer climate in the Kimmeridgian [51]. Sea-level fall is further supported in our study by the co-occurrence of evaporitic minerals floating inside the mudstone and in-situ brecciated mudstones. These observations are consistent with a first positive excursion of the strontium isotope ratio (up to 0.72859) and the heavier oxygen-isotope composition in those samples containing evaporites compared to those of host carbonates free of evaporites. These arguments suggest the beginning of sea-level fall and dry climatic conditions during the deposition of the Barsarin formation in Late-Jurassic times.

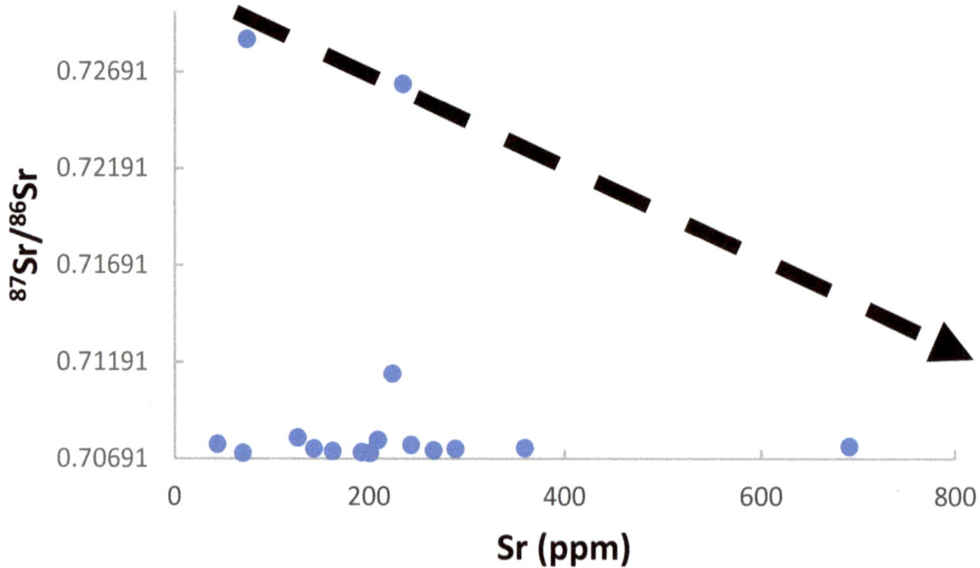

Figure 12. Strontium concentration v. $^{87}Sr/^{86}Sr$ ratios in host carbonate samples; note the variable data of Sr concentration and wide range of Sr isotope ratios.

This discussion can be further evaluated by oxygen and carbon isotopes and petrographic examination: if we considered that high Sr-isotope ratio and low Sr concentrations represent fresh water mixed with evaporated seawater in the host limestones, then $\delta^{13}C$ and $\delta^{18}O$ values would be depleted in their composition. This is the case in Group I, where oxygen-isotope composition shows a wide range, being heavier in the host limestones lacking evaporite. This wide range in Group I suggests the involvement of continental input by a fresh-water river source during the deposition of the Barsarin formation. As a consequence, $\delta^{13}C$ and $\delta^{18}O$ values from the host limestones that closely match the first positive excursion of strontium isotopes could be linked to a short-lived arid episode during Late Jurassic. A global warm climate will increase the rate of chemical weathering [18]. Therefore, warmer climates must increase both riverine Sr flux and its $^{87}Sr/^{86}Sr$ ratio [9]. This is well-supported by the dissolution and brecciation of the mudstones cemented by evaporites (Figure 2C).

The nature of the fluids involved in and after deposition of any carbonate is influenced by the composition of precipitated carbonates, therefore, other environmental conditions need to be considered. The late carbonate cements that contain silica showed relatively high radiogenic signatures (Figures 10 and 11) matching closely the radiogenic isotope values of hot fluids that interacted with the basement or with bedrocks rich in silicate minerals [5,45]. The Sr-isotope values of the dolomite/calcite cements fell into the second positive excursion of $^{87}Sr/^{86}Sr$ values (up to 0.70772) and could be linked in the Barsarin formation to the interaction of hot circulation fluids with the basement recrystallized rocks or interaction of hot fluids with evaporites or siliciclastic rocks (e.g., [52,53]). The negative values of oxygen isotope compositions from dolomite and calcite cements suggest a hot diagenetic fluid involvement [33]. However, the calcite precipitation from meteoric water could also be associated with an enrichment of strontium isotopes and the depletion of $\delta^{13}C$ and $\delta^{18}O$ values, but the negative covariant trend of $\delta^{13}C$ and $\delta^{18}O$ values exclude the involvement of meteoric water [30]. Furthermore, the lower limit of marine isotope concentration is about 200 ppm [54], while the strontium concentration from diagenetic carbonate is populated in two assemblage groups between 148–195 ppm and 362–1029 ppm (Figure 13). The lowest

Sr content, with highest strontium isotopes, indicate a meteoric-derived water, while the highest strontium content with lowest strontium isotopes indicates a possible influence of hot fluid in the diagenetic system. The co-occurrence of dolomite and calcite crystals better supports our scenario, in which dolomite and calcite are attributed to mixing of two different fluids. Therefore, the first positive excursion of direct $^{87}Sr/^{86}Sr$ ratio by laser ablation from the host limestones is partly linked to the weathering riverine strontium input during the depositional time of the Barsarin formation along the ZTFB. The second positive excursion of $^{87}Sr/^{86}Sr$ ratios is most likely related to post-depositional processes, involving two mixed sources of diagenetic fluids in the subsurface setting.

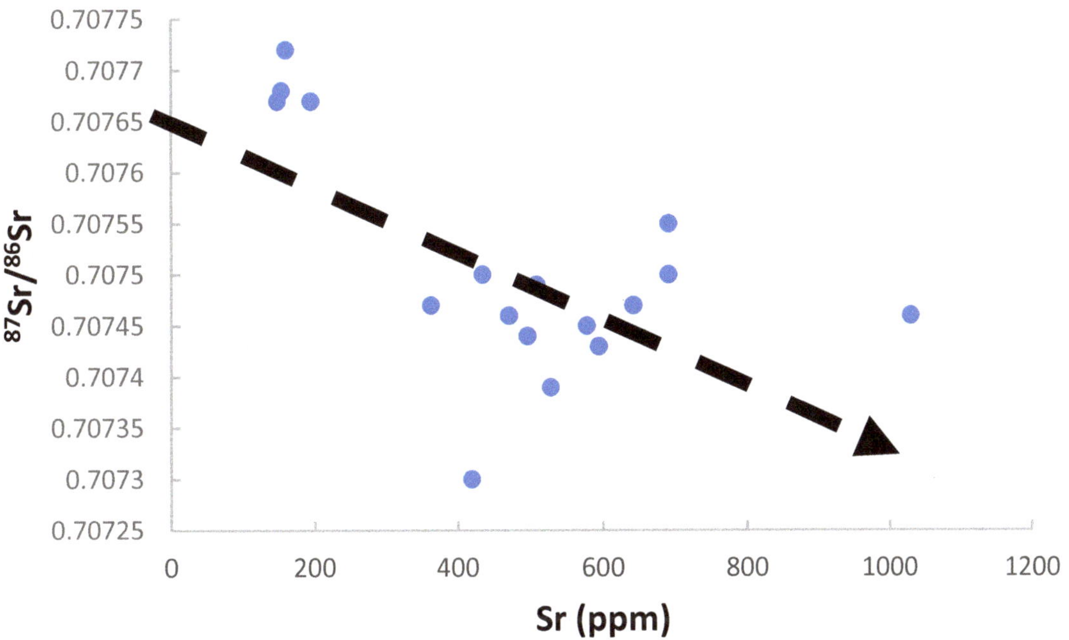

Figure 13. Strontium concentration v. $^{87}Sr/^{86}Sr$ ratios in diagenetic samples, the high strontium concentration is consistent with wide range of $^{87}Sr/^{86}Sr$ ratios, and the low strontium concentration is consistent with high $^{87}Sr/^{86}Sr$ ratios.

6. Conclusions

1. The origin and evolution of host carbonates and the diagenetic fluid history from Barsarin formation were identified on the basis of fieldwork, petrography and geochemistry ($\delta^{13}C$–$\delta^{18}O$ and high resolution $^{87}Sr/^{86}Sr$ isotope ratios by laser ablation on a micrometer-sized scale -MSS).

2. The Barsarin carbonate sequence contains silica infilling the intragranular porosity. Silica is also very abundant, as are inclusions in the dolomite and calcite cements that filled the fractures and pore spaces that affected the formation. The presence of silica indicates that simultaneous diagenetic processes affected the host rocks and cement formation.

3. $\delta^{13}C$–$\delta^{18}O$ isotope values were populated into two groups. The host carbonates of Group I had isotopically heavy oxygen compositions compared to the diagenetic cements of Group II. Group II showed two different assemblages of oxygen-carbon isotope values, suggesting that mixed diagenetic fluids were involved in the precipitation of the carbonate cements.

4. Two positive excursions of the strontium isotope-ratio curve have been identified. The first excursion is related the host carbonates, and the second is the fracture-filling carbonate cements.

5. Two scenarios were proposed for the first positive excursion of the strontium ratios. The first suggests that a silica cycling occurred through dissolution-reprecipitation processes, while the second scenario linked the high values of strontium ratio to weathering of detrital silicate rocks in the headwaters of a fresh river mixed with evaporated seawater. The carbon and oxygen isotopes, and the two assemblage groups of the first excursion of strontium isotopes and strontium contents exclude the first scenario and indicate that the host carbonates formed in a marine evaporative environment mixed with fresh water originated from riverine-rich silicate detrital "Second scenario".

6. The closely associated silica contents with the first positive excursion of absolute $^{87}Sr/^{86}Sr$ ratio in the host carbonates are related to the continuous flux of riverine input during Late Jurassic.

7. Silica and evaporites with low Sr concentration, high $^{87}Sr/^{86}Sr$ ratios and wide range of oxygen and carbon isotope compositions point to a riverine weathering input in the evaporated seawater during the sea level fall which occurred during the Scenario II event.

8. The wide range of $\delta^{13}C$–$\delta^{18}O$ values in Group II and their positive covariance suggest a late fluid mixing during the second excursion of the $^{87}Sr/^{86}Sr$ ratio. In this case mixed fluids, consisting probably of low-temperature meteoric and hot diagenetic fluids, were inferred from the stable isotope compositions of the post-depositional carbonates "calcite and dolomite cements".

Author Contributions: Writing and preparation the original draft, N.S. and A.P.; review and editing, K.K., J.-N.P., N.S. and A.P., Methodology and software, A.G., A.P. and N.S. All authors have read and agreed to the published version of the manuscript.

Funding: The study benefited from research funds of the Université Libre de Bruxelles (ULB)-Belgium.

Acknowledgments: This work was supported by Université Libre de Bruxelles-Belgium and University of Alberta-Canada.

Conflicts of Interest: The authors declare no conflict of interest.

References

1. Fox, J.; Ahlbrandt, E.; Thomas, S. *Petroleum Geology and Total Petroleum Systems of the Widyan Basin and Interior Platform of Saudi Arabia and Iraq*; Bulletin 2202-E; U.S. Geological Survey: Reston, VA, USA, 2002; Volume 26. [CrossRef]
2. Bellen, R.C.V.; Dunnington, H.V.; Wetzel, R.; Morton, D. Lexique Stratigraphic International: Fascicule 10a, Asie, Iraq. In *Mesozoic and Palaeozoic*; van Bellen, R.C., Ed.; Centre National de la Recherche Scientifique: Paris, France, 1959; Volume 3.
3. Maliva, R.G.; Siever, R. Nodular Chert Formation in Carbonate Rocks. *J. Geol.* **1989**, *97*, 421–433. [CrossRef]
4. Gao, G.; Land, L. Nodular chert from the Arbuckle Group, Slick Hills, SW Oklahoma: A combined field, petrographic and isotopic study. *Sedimentology* **1991**, *38*, 857–870. [CrossRef]
5. Salih, N.; Mansurbeg, H.; Préat, A. Geochemical and Dynamic Model of Repeated Hydrothermal Injections in Two Mesozoic Successions, Provençal Domain, Maritime Alps, SE-France. *Minerals* **2020**, *10*, 775. [CrossRef]
6. Fairchild, I.J.; Smith, C.L.; Baker, A.; Fuller, L.; Spötl, C.; Mattey, D.; McDermott, F. Modification and preservation of environmental signals in speleothems. *Earth Sci. Rev.* **2006**, *75*, 105–153. [CrossRef]
7. Tullborg, E.L.; Drake, H.; Sandstrom, B. Palaeohydrogeology: A methodology based on fracture mineral studies. *Appl. Geochem.* **2008**, *23*, 1881–1897. [CrossRef]
8. Salih, N.; Mansurbeg, H.; Kolo, K.; Préat, A. Hydrothermal Carbonate Mineralization, Calcretization, and Microbial Diagenesis Associated with Multiple Sedimentary Phases in the Upper Cretaceous Bekhme Formation, Kurdistan Region-Iraq. *Geosciences* **2019**, *9*, 459. [CrossRef]
9. Jones, E.; Jenkyns, C. Seawater strontium isotopes, oceanic anoxic events, and seafloor hydrothermal activity in the Jurassic and Cretaceous. *Am. J. Sci.* **2001**, *301*, 112–149. [CrossRef]
10. Mountjoy, E.W.; Machel, H.G.; Green, D.; Duggan, J.; Williams-Jones, A.E. Devonian matrix dolomites and deep burial carbonate cements: A comparison between the Rimbey-Meadowbrook reef trend and the deep basin of west-central Alberta. *Bull. Can. Petrol. Geol.* **1999**, *47*, 487–509.

11. Jones, C.E.; Jenkyns, H.C.; Coe, A.L.; Hesselbo, S.P. Strontium isotopic variations in Jurassic and Cretaceous seawater. *Geochim. Cosmochim. Acta* **1994**, *58*, 3061–3074. [CrossRef]
12. Jassim, S.Z.; Goff, J.C. *Geology of Iraq*; Dolin, Prague and Moravian Museum: Brno, Czech Republic, 2006; 341p.
13. Buday, T. The Regional Geology of Iraq. In *Stratigraphy and Paleogeography*; Kassab, I.I., Jassim, S.Z., Eds.; GEOSURV: Baghdad, Iraq, 1980; Volume 1, 445p.
14. Rankenburg, K.; Lassiter, J.C.; Brey, G. Origin of megacrysts in volcanic rocks of the Cameroon volcanic chain—Constraints on magma genesis and crustal contamination. *Contrib. Mineral. Petrol.* **2004**, *147*, 129–144. [CrossRef]
15. Mamet, B.; Préat, A. Jurassic microfacies, Rosso Ammonitico limestone, Subbetic Cordillera, Spain. *Rev. Esp. Micropaleontol.* **2006**, *38*, 219–228.
16. Brand, U. Aragonite-calcite transformation based on Pennsylvanian molluscs. *Geol. Soc. Am. Bull.* **1989**, *101*, 377–390. [CrossRef]
17. Anderson, T.F.; Popp, B.N.; Williams, A.C.; Ho, L.-Z.; Hudson, J.D. The stable isotopic record of fossils from the Peterborough Member, Oxford Clay Formation (Jurassic), UK: Paleoenvironmental implications. *J. Geol. Soc. Lond.* **1994**, *151*, 125–138. [CrossRef]
18. Veizer, J.; Ala, D.; Azmy, K.; Bruckschen, P.; Buhl, D.; Bruhn, F.; Carden, G.A.F.; Diener, A.; Ebneth, S.; Godderis, Y.; et al. 87Sr/86Sr, δ13C and δ18O Evolution of Phanerozoic Seawater. *Chem. Geol.* **1999**, *161*, 59–88. [CrossRef]
19. Jenkyns, H.C.; Schouten-Huibers, L.; Schouten, S.; Sinninghe Damsté, J.S. Warm Middle Jurassic–Early Cretaceous high-latitude sea-surface temperatures from the Southern Ocean. *Clim. Past* **2012**, *8*, 215–226. [CrossRef]
20. Price, G.D.; Sellwood, B.W. Paleotemperature indicated by upper Jurassic (Kimmeridgian Tithonian) fossils from Mallorca determined by oxygen-isotope composition. *Palaeogeogr. Palaeoclimatol. Palaeoecol.* **1994**, *110*, 1–10. [CrossRef]
21. Madhavaraju, J.; Sial, A.N.; González-León, C.M.; Nagarajan, R. Carbon and oxygen isotopic variations in early Albian limestone facies of the Mural Formation, Pitaycachi section, northeastern Sonora, Mexico. *Rev. Mex. Cienc. Geológi.* **2013**, *30*, 526–539.
22. Tanner, L.H. Continental Carbonates as Indicators of Paleoclimate. In *Carbonates in Continental Settings: Geochemistry, Diagenesis and Applications*; Developments in Sedimentology Series, Alonso-Zarza, A.M., Tanner, L.H., Eds.; Elsevier: Amsterdam, The Netherlands, 2010; Volume 62, pp. 179–214. [CrossRef]
23. Grocke, D.R.; Hori, R.S.; Arthur, M.A. The global significance of a deep-sea isotopic event during the Toarcian oceanic anoxic event recorded in Japan. *Am. Geophys. Union* **2003**, *84*, 905.
24. Li, M.; Fang, X.; Li, J.; Yan, M.; Sun, S.; Zhu, L. Isotopic Application in High Saline Conditions. In *Isotopes Applications in Earth Sciences*; Abdel Rahman, R.O., Ed.; IntechOpen: Rijeka, Croatia, 2020; Chapter 3. [CrossRef]
25. Magaritz, M.; Stemmerik, L. Oscillation of carbon and oxygen isotope compositions of carbonate rocks between evaporative and open marine environments, Upper Permian of East Greenland. *Earth Planet. Sci. Lett.* **1989**, *93*, 233–240. [CrossRef]
26. Warfe, D.; Pettit, N.; Davies, P.; Pusey, B.; Hamilton, S.; Kennard, M.; Townsend, S.; Bayliss, P.; Ward, D.; Douglas, M. The 'wet–dry' in the wet–dry tropics drives river ecosystem structure and processes in northern Australia. *Fresh Water Biol.* **2011**, *56*, 2169–2195. [CrossRef]
27. Lécuyer, C.; Allemand, P. Modelling of oxygen isotope evolution of seawater: Implications for the climate interpretation of the $\delta^{18}O$ of marine sediments. *Geochim. Cosmochim. Acta* **1999**, *63*, 351–361. [CrossRef]
28. Vickers, L.M.; Bajnai, D.; Price, G.D.; Linckens, J.; Fiebig, J. Southern high-latitude warmth during the Jurassic–Cretaceous: New evidence from clumped isotope thermometry. *Geology* **2019**, *47*, 724–728. [CrossRef]
29. Padden, M.; Weissert, H.; Funk, H.; Schneider, S.; Gansner, C. Late Jurassic lithological evolution and carbon isotope stratigraphy of the Western Tethys. *Eclogae Geol. Helv.* **2002**, *95*, 333–346.
30. Deocampo, M. The Geochemistry of Continental Carbonates. In *Carbonates in Continental Settings: Geochemistry, Diagenesis and Applications*; Developments in Sedimentology Series, Alonso-Zarza, A.M., Tanner, L.H., Eds.; Elsevier: Amsterdam, The Netherlands, 2010; Volume 62, pp. 1–59.
31. Zheng, Y.F.; Hoefs, J. Carbon and oxygen covariations in hydrothermal calcites: Theoretical modelling on mixing processes and application to Pb-Zn deposits in the Harz Mountains. *Miner. Depos.* **1993**, *28*, 79–89. [CrossRef]
32. Moore, H. Carbonate Diagenesis and Porosity. *Dev. Sedimentol.* **1989**, *46*, 338.
33. Salih, N.; Mansurbeg, H.; Kolo, K.; Gerdes, A.; Préat, A. In situ U-Pb dating of hydrothermal diagenesis in tectonically controlled fracturing in the Upper Cretaceous Bekhme Formation, Kurdistan Region-Iraq. *Int. Geol. Rev.* **2020**, *62*, 2261–2279. [CrossRef]
34. Bustillo, M.A. Silicification of Continental Carbonates. In *Carbonates in Continental Settings: Processes, Facies and Applications*; Developments in Sedimentology Series, Alonso-Zarza, A.M., Tanner, L.H., Eds.; Elsevier: Amsterdam, The Netherlands, 2010; Volume 62, pp. 153–174.
35. Bustillo, M.A.; Pérez-Jiménez, J.L.; Alonso-Zarza, A.M.; Furio, M. Moganite in the chalcedony varieties of continental cherts (Miocene, Madrid basin, Spain). *Spectrosc. Lett.* **2012**, *45*, 109–113. [CrossRef]
36. McArthur, M.; Howarth, J.; Shields, A. Strontium Isotope Stratigraphy. In *The Geologic Time Scale*; Gradstein, F.M., Ogg, J.G., Schmitz, M., Ogg, G.M., Eds.; Elsevier: Boston, MA, USA, 2012; Volume 2, pp. 127–144.
37. Korte, C.; Kozur, W.; Bruckschen, P.; Veizer, J. Strontium isotope evolution of Late Permian and Triassic seawater. *Geochim. Cosmochim. Acta* **2003**, *67*, 47–62. [CrossRef]
38. Cantarero, I.; Parcerisa, D.; Plata, M.A.; Gómez-Gras, D.; Gomez-Rivas, E.; Martín-Martín, J.D.; Travé, A. Fracturing and Near-Surface Diagenesis of a Silicified Miocene Deltaic Sequence: The Montjuïc Hill (Barcelona). *Minerals* **2020**, *10*, 135. [CrossRef]

39. Shen, B.; Ye, H.; Ma, H.; Lang, X.; Pei, H.; Zhou, C.; Zhang, S.; Yang, R. Hydrothermal origin of sydepositional chert bands and nodules in the Mesoproterozoic Wumishan Formation: Implication for the evolution of Mesoproterozoic cratonic basin, North China. *Precambrian Res.* **2018**, *310*, 213–228. [CrossRef]
40. Wen, H.; Fan, H.; Tian, S.; Wang, Q.; Hu, R. The formation conditions of the early Ediacaran cherts, South China. *Chem. Geol.* **2016**, *430*, 45–69. [CrossRef]
41. Palmer, M.R.; Edmond, J.M. Controls over the strontium isotope composition of river water. *Geochim. Cosmochim. Acta* **1992**, *56*, 2099–2111. [CrossRef]
42. Wierzbowski, H.; Anczkiewicz, R.; Pawlak, J.; Rogov, M.A.; Kuznetsov, A.B. Revised Middle—Upper Jurassic Strontium Isotope Stratigraphy. *Chem. Geol.* **2017**, *466*, 239–255. [CrossRef]
43. Mohajjel, M.; Fergusson, C.L. Jurassic to Cenozoic tectonics of the Zagros Orogen in northwestern Iran. *Int. Geol. Rev.* **2014**, *56*, 263–287. [CrossRef]
44. Lechmann, A.; Burg, J.-P.; Ulmer, P.; Guillong, M.; Faridi, M. Metasomatized mantle as the source of Mid-Miocene-Quaternary volcanism in NW-Iranian Azerbaijan: Geochronological and geochemical evidence. *Lithos* **2018**, *304–307*, 311–328. [CrossRef]
45. Davies, G.R.; Smith, L.B. Structurally controlled hydrothermal dolomite reservoir facies: An overview. *AAPG* **2006**, *90*, 1641–1690. [CrossRef]
46. Hodell, D.A.; Mead, G.A.; Mueller, P.A. Variation in the strontium isotopic composition of seawater (8 Ma to present): Implications for chemical weathering rates and dissolved fluxes to the oceans. *Chem. Geol.* **1990**, *80*, 291–307. [CrossRef]
47. Matyszkiewicz, J. Thz significance of Saccocoma-calciturbidites for the analysis of the Polish Epicontinetal Late Jurassic Basin: An example from the Southern Cracow-Wielun Upland (Poland). *Facies* **1996**, *34*, 23–40. [CrossRef]
48. Breesch, L.; Swennen, R.; Vincent, B.; Ellison, R.; Dewever, B. Dolomite cementation and recrystallisation of sedimentary breccias along the Musandam Platform margin (United Arab Emirates). *J. Geochem. Explor.* **2010**, *106*, 34–43. [CrossRef]
49. Burke, W.H.; Denison, R.E.; Heatherington, E.A.; Koepnick, R.B.; Nelson, H.F.; Otto, J.B. Variation of seawater $^{87}Sr/^{86}Sr$ through Phanerozoic time. *Geology* **1982**, *10*, 516–519. [CrossRef]
50. Zheng, Y.F.; Chen, J.F. *Stable Isotope Geochemistry*; Science Press: Beijing, China, 2000. (In Chinese)
51. Haq, B.U. Jurassic sea-level variations: A reappraisal. *GSA Today* **2018**, *28*, 4–10. [CrossRef]
52. Wendte, J.; Dravies, J.; Stasiuk, D.; Qing, H.; Moore, O.; Ward, G. High temperature saline (thermoflux) dolomitization of Devonian Swan Hills platfoand bank carbonates, Wild River area, west-central Alberta. *Bull. Can. Pet. Geol.* **1998**, *46*, 210–265.
53. Spencer, J.; Jeary, V.; Moore, O. Sedimentary Exhalative Dolomite from the Middle Cambrian Eldon and Cathedral Formations of the Canadian Rocky Mountains. In *Dolomites—The Spectrum: Mechanisms, Models, Reservoir Development*; Extended Abstracts, McAuley, R., Eds.; Canadian Society of Petroleum Geologists, Seminar and Core Conference: Calgary, AB, Canada, 2004; pp. 13–15.
54. Derry, L.; Keto, L.S.; Jacobsen, S.B.; Knoll, A.H.; Swett, K. Sr isotopic variations in Upper Proterozoic carbonates from Svalbard and East Greenland. *Geochim. Cosmochim. Acta* **1989**, *53*, 2331–2339. [CrossRef]

Article

Dolomitization of Paleozoic Successions, Huron Domain of Southern Ontario, Canada: Fluid Flow and Dolomite Evolution

Ihsan S. Al-Aasm [1,*], Richard Crowe [2] and Marco Tortola [1]

1 School of the Environment, University of Windsor, 401 Sunset Avenue, Windsor, ON N9B 3P4, Canada; tortolam@uwindsor.ca
2 Nuclear Waste Management Organization, 22 St. Clair Avenue, East, Toronto, ON M4T 2S3, Canada; rcrowe@nwmo.ca
* Correspondence: alaasm@uwindsor.ca

Abstract: Integrated petrographic, isotopic, fluid inclusion microthermometry, and geochemical analyses of Paleozoic carbonate successions from multiple boreholes within the Huron Domain, southern Ontario were conducted to characterize the diagenetic history and fluid composition, on a regional scale, and evaluate the nature and origin of dolomitized beds. Multiple generations of non-stochiometric dolomite have been observed. These dolomites occur as both replacement (D1 and D2) and cement (saddle dolomite; SD) and formed either at near-surface to shallow burial zone (D1) or intermediate burial (D2 and SD). Petrographic and geochemical data of dolomite types and calcite cement suggest that these carbonates have experienced multiple fluid events that affected dolomite formation and other diagenetic processes. Cambrian and Ordovician strata have two possibly isolated diagenetic fluid systems; an earlier fluid system that is characterized by a pronounced negative shift in oxygen and carbon isotopic composition, more radiogenic Sr ratios, warm and saline signatures, higher average $\sum$REE compared to warm water marine brachiopods, negative La anomaly, and positive Ce anomaly; and a later Ordovician system, characterized by less negative shifts in oxygen and carbon isotopes, comparable Th, hypersaline, a less radiogenic, less negative La anomaly, and primarily positive Ce anomaly but also higher average $\sum$REE compared to warm water marine brachiopods. Ordovician, Silurian, and Devonian Sr isotopic ratios, however, show seawater composition of their respective age as the primary source of diagenetic fluids with minor rock/water interactions. In contrast, the isotopic data of the overlying Silurian and Devonian carbonates show overlaps between δ^{13}C and δ^{18}O values. However, δ^{18}O values show evidence of dolomite recrystallization. D2 shows wide T_h values and medium to high salinity values. Higher Th and salinity are observed in SD in the Silurian carbonates, which suggest the involvement of localized fluxes of hydrothermal fluids during its formation during Paleozoic orogenesis. Geochemical proxies suggest that in both age groups the diagenetic fluids were originally of coeval seawater composition, subsequently modified via water-rock interaction possibly related to brines, which were modified by the dissolution of Silurian evaporites from the Salina series. The integration of the obtained data in the present study demonstrates the linkage between fluid flux history, fluid compartmentalization, and related diagenesis during the regional tectonic evolution of the Michigan Basin.

Keywords: fluid flow; dolomitization; Paleozoic; Huron Domain

Citation: Al-Aasm, I.S.; Crowe, R.; Tortola, M. Dolomitization of Paleozoic Successions, Huron Domain of Southern Ontario, Canada: Fluid Flow and Dolomite Evolution. *Water* **2021**, *13*, 2449. https://doi.org/10.3390/w13172449

Received: 3 August 2021
Accepted: 30 August 2021
Published: 6 September 2021

1. Introduction

As part of an ongoing investigation into the occurrence of strata-bound dolomite layers within sedimentary, Paleozoic-age formations of southern Ontario an initial study using core acquired from the site of a formerly proposed deep geological repository (Bruce nuclear site) was conducted [1]. This initial study provided a site-specific analog to examine fluid migration and rock-formation barrier integrity at geological timescales. Results showed that hydrothermal alterations were minimal, burial-related, and lacked isotopic signatures associated with fault-controlled diagenesis, as observed in other regions of southern Ontario. This program was then expanded to a regional scale, to determine if the conditions observed at the Bruce nuclear site were consistent across the Huron Domain of southern Ontario. Core samples from multiple deep boreholes within the Huron Domain of southern Ontario were analyzed for petrographic, stable, and Sr isotopic composition, fluid inclusion microthermometry, and major, trace, and REE to characterize diagenetic history, fluid composition, and sedimentary provenance (e.g., [2,3]).

Extensive fluid flow occurs during tectonic thrusting, sedimentary loading, uplift, and compression in many sedimentary basins [4]. These fluids can affect the redistribution of mass, heat, petroleum migration, and formation of sediment-hosted ore deposits [5–7]. A fundamental problem in geology is to constrain the composition of the fluids within the context of thermal and tectonic evolution (i.e., changes of stress regimes) of hosting sedimentary basins. The composition and evolution of ancient sedimentary fluids have been successfully reconstructed using recent advances in stable and radiogenic isotopes, trace and rare-earth element geochemistry, fluid inclusion analyses, and paleomagnetic techniques (e.g., [7–13]). The fluids identified in these studies ranged in composition from evaporative brines, marine, mixed marine-meteoric, to deep brines and hydrothermal waters.

Dong et al. [14] showed that strata-bound dolostones form in a broad range of geologic and geochemical processes and generally result from large-scale fluid flow. Important aspects of dolomitization that need further thorough investigation are: (1) the source and supply of Mg [15,16]; (2) the plumbing mechanism necessary to transport Mg-bearing solutions to the site of dolomitization [17]; (3) the stoichiometry and recrystallization of dolomite [18–23], and (4) the effect of tectonics on the fluid flow [5,24–27]. An understanding of the genesis of ancient dolomites is only possible through detailed petrographic and geochemical studies of dolomitized sequences where the geologic history of the rocks is also reasonably well constrained.

There are several models suggested explaining the formation of dolomite over a broad range of geologic and geochemical processes. These models include [28–31]: (1) sabkha dolomitization model, (2) seepage-reflux dolomitization model, (3) mixing zone dolomitization model, (4) seawater dolomitization model (5) burial-dolomitization model, and (6) structurally-controlled hydrothermal dolomitization model.

Geological and geochemical processes lead to a change in formation density that can alter the physical properties of the bedrock, including increased porosity and permeability. In this sense, dolomitization can influence groundwater system evolution through the creation of permeable pathways following lithification, especially along localized geologic structures suitable for the accumulation of hydrocarbons. These models are dependent on three criteria [30]: (1) Thermodynamic: the fluids must be supersaturated with respect to dolomite and undersaturated with respect to calcite. Thus, dolomite cement will precipitate, and calcite cement will dissolve. (2) Kinetic: the rate of dolomite precipitation must be equal to or greater than the rate of calcite dissolution. (3) Hydrologic: Mg-rich fluids must be pumped continuously in the limestone pore system, faults, and fractures [30].

Recent studies by Haeri–Ardakani et al. [32,33] identified three distinct dolomite types within southwestern Ontario: a microcrystalline dolomite (D1) formed during early diagenesis from Middle Ordovician seawater and recrystallized during progressive burial, a medium crystalline dolomite (D2) formed by hydrothermal fluids (68 to 99 °C) and a late-stage saddle dolomite cement (D3) related to fault controlled, high temperature fluids (>125 °C).

Most recently, Al-Aasm [1], Al-Aasm and Crowe [2] and Tortola et al. [3] assessed the nature and origin of strata bound dolomitization occurring in rock samples from Cambrian through Devonian carbonates and siliciclastics beneath the Bruce nuclear site and adjacent areas within southern Ontario. These latter studies better supported the notion of fluid compartmentalization between different stratigraphic successions.

Hence, the main purpose of this study is to thoroughly examine all data gathered in both previous and new studies by the authors to further evaluate, on a regional scale, the nature and origin of dolomitized beds within Paleozoic carbonates of the Huron Domain with the intention on deciphering the nature of dolomitization processes, evolution of diagenetic fluids, the nature of paleohydrogeological system(s) and whether they are connected on not, the source of heat for hydrothermal fluids (if any) and the driving force for diagenetic fluid(s) migration, in contrast to previous studies in southwestern Ontario.

2. Regional Geologic, Tectonics, and Stratigraphic Framework

The study area is located on the northeastern margin of the Michigan Basin (Figures 1 and 2). It represents a portion of the northwestern flank of the Algonquin Arch that consists of Paleozoic sediments overlying the subsurface basement [34]. The thickness of the Paleozoic sequence is estimated to range from a maximum of 4800 m at the center of the Michigan Basin to 850 m on the flank of the Algonquin Arch.

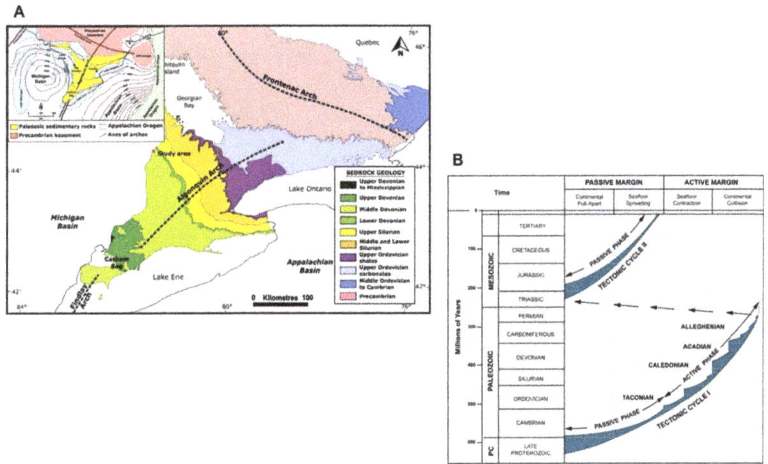

Figure 1. (**A**) Generalized Paleozoic bedrock geology of southern Ontario (modified from [35,36]); (**B**) Tectonic history of the Phanerozoic successions in southern Ontario (modified from [37]).

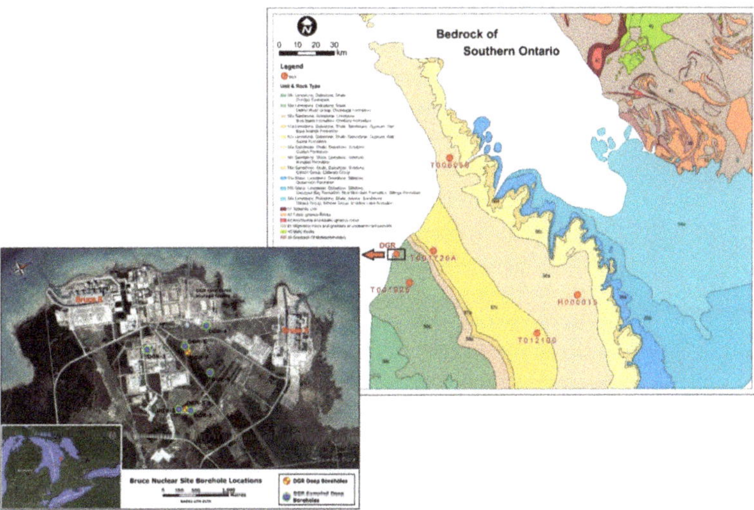

Figure 2. Bedrock geology of southern Ontario showing the location of boreholes used for this study within the Bruce Nuclear site and other areas in southern Ontario (modified from [38]).

Southern Ontario separates two major Paleozoic sedimentary basins, the Appalachian Basin to the southeast and Michigan Basin to the west (Figure 1A). The Algonquin Arch, which separates the two basins, extends from northeast of the Canadian shield to the southwest near the city of Chatham. Near the Windsor area, another structural high named the Findlay Arch begins and extends to the southwest into Michigan and Ohio states [39]. The Michigan basin is a nearly circular intracratonic basin dominated by carbonate and evaporite sediments, whereas the Appalachian Basin is an elongate foreland basin developed because of the collisional tectonic event along the eastern margin of the North American continent during the Paleozoic and is mainly characterized by deposition of siliciclastic sediments [40].

During Precambrian to Cambrian, a rifting event that separated North America and Africa initiated subsidence (approximately 1100 Ma) and deposition within the Michigan Basin (Figure 1B; [37]). Thermal subsidence of the Precambrian basement followed approximately 580 Ma to 500 Ma after the cooled lithosphere increased in density [41]. The subsequent and continuous deposition of sediment into the Michigan Basin caused the bending of the basement [42–44]. However, the development of the basin is not the result of a continuous subsidence but represents the outcome of a series of tectonic events that characterized the Paleozoic [45,46].

The Taconic and Acadian Orogenies (Figure 1B) had a dominant control on the Paleozoic strata of the eastern flank of Michigan [47]. In fact, during the Taconic Orogeny (Ordovician), a large-scale eastward tilting of the Laurentian margin has been produced with the consequent disappearance of the concentric depositional geometry of the Michigan Basin and with the subsequent deposition of Ordovician limestone and overlying shale successions [46,48,49].

The Middle Devonian Acadian Orogeny (Figure 1B) determined the reactivation of the Algonquin-Findlay Arch system with consequent return of the concentric geometry of the Michigan Basin [45]. During the Upper Silurian where restricted marine conditions dominated, the deposition was mainly characterized by the evaporites of the Salina Group [50,51]. In addition, the presence of an unconformity reveals emergent conditions at the end of the Silurian [47].

The Caledonian (Devonian) and Alleghenian (Carboniferous) Orogenies (Figure 1B) played a major role in terms of diagenetic fluid migration [5]. The regional maximum principal stress at that time was northwesterly oriented in accordance with the main direction of the thrust motion along the Appalachian tectonic front [47,52].

The Paleozoic sequence is separated from underlying Precambrian rocks of the ca. 1.1 Ga old Grenville orogeny by an unconformity overlain by a relatively thin Cambrian sandstone, which represents an aquifer [53,54]. The depositional environment of the Ordovician and Cambrian formations encompasses various facies assemblages, including supratidal, lagoonal to subtidal environments (e.g., [55,56]). The diverse facies changes observed in these formations were controlled by relative sea-level changes and sources of detritus, related to the regional tectonic setting during the Paleozoic [37,46]. In the study area, many of the Cambrian units along the Algonquin Arch (Figure 1) were absent due to arch rejuvenation and uplift during early Paleozoic times (e.g., [56]). Ordovician formations consist of thin, decimeter-scale, alternating shale, and carbonate beds; some of which have been dolomitized [47]. The Michigan Basin has been characterized by a typical tropical climate during mid to late Silurian because it was located at 25° south of the equator [57] leading to the development of shallow waters reef complexes. Moreover, the Michigan Basin has been isolated and is becoming an evaporative basin in which alternation of evaporites and carbonates is deposited during the rest of the Silurian [58]. Conversely, in the Middle Devonian, the Michigan basin has been characterized by an extremely arid climate with consequent deposition of shallow intertidal and supratidal lithofacies such as Amherstburg and Lucas formations [59]. The Devonian sequences are represented by limestones, shales, dolostones, and sandstones [39] with arid climate and deposition of shallow intertidal and supratidal lithofacies (e.g., Amherstburg and Lucas formations).

3. Sampling and Analytical Methods

This section may be divided into subheadings. It should provide a concise and precise description of the experimental results, their interpretation, as well as the experimental conclusions that can be drawn.

Cores (n = 91) from specific sections from deep boreholes (n = 10) from Bruce Nuclear Site (Tiverton, Ontario) and across the Huron Domain of southern Ontario were sampled (Figure 2). These samples encompass formations ranging from Cambrian to Devonian age successions and representing a wide range of rocks including dolomitized limestones, dolostones, sandy dolostones, and sandstones (Figure 2). These core samples were examined and photographed prior to their selection for making thin sections and fluid inclusion wafers. Petrographic examination of 145 thin sections was performed under a standard petrographic microscope to compliment core descriptions and by cathodoluminescence microscopy (CL) using a Technosyn cold cathodoluminescence stage with a 12–15 kV beam and a current intensity of 0.42–0.43 mA.

Samples for oxygen and carbon isotopic analyses were extracted from calcite and dolomite samples (n = 259) from polished slabs using a microscope-mounted drill assembly. The samples were reacted in a vacuum with 100% pure phosphoric acid for four hours at 25° and 50 °C for calcite and dolomite, respectively. CO_2 gas from samples that contained a mixture of calcite and dolomite was extracted using the chemical separation method of Al-Aasm et al. [60]. The evolved CO_2 gas was analyzed for isotopic ratios on a Delta Plus mass spectrometer. Delta (δ) values for oxygen and carbon are reported per mil (‰) relative to the PeeDee Belemnite (VPDB) standard. Precision was better than 0.05‰ for both $\delta^{18}O$ and $\delta^{13}C$.

Twenty-six core samples were selected for Sr isotopic analyses from various calcitic and dolomitic components representing host rock, skeletal components, and diagenetic phases. Prior to $^{87}Sr/^{86}Sr$ analysis, powdered samples (n = 78) were extracted employing a microscope-mounted drill assembly to extract 1–10 mg of powdered calcite or dolomite from polished slabs. Care was taken to avoid contamination from other components such as later cement fill. These samples were initially dissolved in 2.5 N HCl suprapure for 25 h in sealed PFA vessels at room temperature. They were then passed through cation exchange quartz columns. To remove the eventual pore salts that may contain Sr, these samples were, after crushing and pulverization were washed twice with MilliQ water. Strontium isotopic ratios were measured on a Finnigan MAT 261 mass spectrometer. All analyses were performed in the static multi-collector mode using Re filaments. NBS and ocean water were used as standard references and $^{87}Sr/^{86}Sr$ ratios were normalized to $^{87}Sr/^{86}Sr$ = 8.375209. The mean standard error was 0.00003 for NBS-987.

Thirty-one samples were analyzed for fluid inclusion microthermometric studies. Petrographic work was done with a conventional petrographic microscope. Careful and detailed observations of size, occurrence habit, and paragenetic contexts of the fluid inclusions were conducted in the double polished wafers. Distinctions were made between primary, secondary, and pseudo-secondary fluid inclusions. This was followed by a determination of the number of phases (i.e., the liquid to vapor ratio) and their ratios entrapped in each of the fluid inclusions. Then determination of the composition (aqueous or hydrocarbon fluids) of the inclusion using fluorescence microscopy; most oil-filled inclusions fluoresce, while the aqueous inclusions do not. Finally, microthermometric measurements of the homogenization (Th) and final ice melting temperatures (Tmice) of fluid inclusions were measured. These measurements were conducted using a Linkam THMGS600 heating-freezing stage, which was calibrated using synthetic fluid inclusions of known compositions. Homogenization temperatures (Th), and ice-melting temperatures (Tm-ice) were measured with a precision (reproducibility) of ±1 °C and ±0.1 °C, respectively. The salinities for inclusions in which ice was the last melting phase were calculated using the program of Chi and Ni [61], and those for inclusions with hydrohalite as the last melting phase was done with the program of Steele-MacInnis et al. [62].

Deformation of inclusion walls, necking-down of single inclusions into smaller inclusions, or leakage of fluid can all lead to changes in homogenization temperature. To limit the possibility of measuring deformed aqueous inclusions, only primary inclusions from the same field of view were measured during a single heating or freezing run. By restricting measurements to inclusions within the same field of view, any sudden changes in liquid/vapor ratios due to inclusion deformation could be observed and removed from consideration. This does not exclude the possibility of subtle deformation in the z-axis, which is difficult to detect. Heating runs were conducted before freezing runs to reduce the possibility of inclusion stretching by freezing [63].

Major (Ca, Mg), minor, trace elements (Sr, Na, Mn, and Fe) and rare-earth element (REE) analyses (n = 95) were performed using ICP-MS. The procedure used for the sample digestion has been adapted to digest carbonates only. Around 50–100 mg of powdered carbonate samples from different calcite and dolomite phases were digested utilizing 5% v/v acetic acid. Measurements were carried employing an Agilent 8800 inductively coupled mass spectrometer at the University of Waterloo. All results we reported in part per million (ppm).

For rare-earth (REE) element analysis, samples were analyzed using external calibration, internal standard, and standard addition methods, following the procedure described in Jenner et al. [64]. Measured REE values were normalized to Post Archean Australian Shale (PAAS). Rare-earth element anomalies such as Ce_{SN} [$(Ce/Ce^*)_{SN} = Ce_{SN}/(0.5La_{SN} + 0.5Pr_{SN})$], La_{SN} [$(Pr/Pr^*)_{SN} = Pr_{SN}/0.5Ce_{SN} + 0.5Nd_{SN})$], Eu_{SN} [$(Eu/Eu^*)_{SN} = Eu_{SN}/(0.5Sm_{SN} + 0.5Gd_{SN})$], and Gd_{SN} [$(Gd/Gd^*)_{SN} = Gd_{SN}/(0.33Sm_{SN} + 0.67Tb_{SN})$] were calculated employing the equations proposed by Bau and Dulski [65]. The proportions of LREE(La-Nd) over HREE (Ho-Lu) were calculated via $(La/Yb)_{SN}$ ratios as proposed by Kucera et al. [66]. The proportions of MREE (Sm-Dy) was calculated using $(Sm/La)_{SN}$ and $(Sm/Yb)_{SN}$ ratios [67] over LREE and HREE, respectively.

An environmental scanning electron microscope (ESEM) coupled with an EDAX detector was utilized to analyze selected carbon-coated thin section samples to investigate the nature of dolomitization and subsequent recrystallization.

4. Results

4.1. Petrographic Analysis

Cambrian strata are dominated by mixed carbonate-siliciclastic lithofacies (Figure 3). The carbonate lithofacies is composed of dolomitized and silicified crinoidal and bryozoan components. The siliciclastic facies consist of silicified ooids and other coated grains as well as detrital carbonates with large detrital quartz, cemented by carbonates.

The Middle Ordovician carbonates in the subsurface are divided into the Black River Group and the overlying Trenton Group ([47]; Figure 3). The original depositional facies of the investigated Ordovician formations (Shadow Lake, Gull River, Coboconk, Kirkfield, Sherman Fall, Cobourg, and Collingwood) consists of nodular, bioturbated, micritic, and bioclastic facies—now partly dolomitized. In the Coboconk Formation, a thin fossiliferous layer (grainstone facies) occurs with a sharp contact with the nodular facies. This layer is also pervasively dolomitized. Skeletal components include bryozoans and echinoderms. In the Gull River Formation, the matrix is highly dolomitized but fossils such as brachiopods, crinoids, and trilobites are dolomitized but their textural integrity is preserved.

The Silurian formations in the Huron Domain are characterized by several lithofacies of partially to fully dolomitized carbonates, such as dolommudstone with abundant fenestrae, sometimes brecciated, and coated and bioclastic grainstone lithofacies. These carbonate lithofacies alternate with evaporite units, such as A-1, A-2, and B (Figure 3).

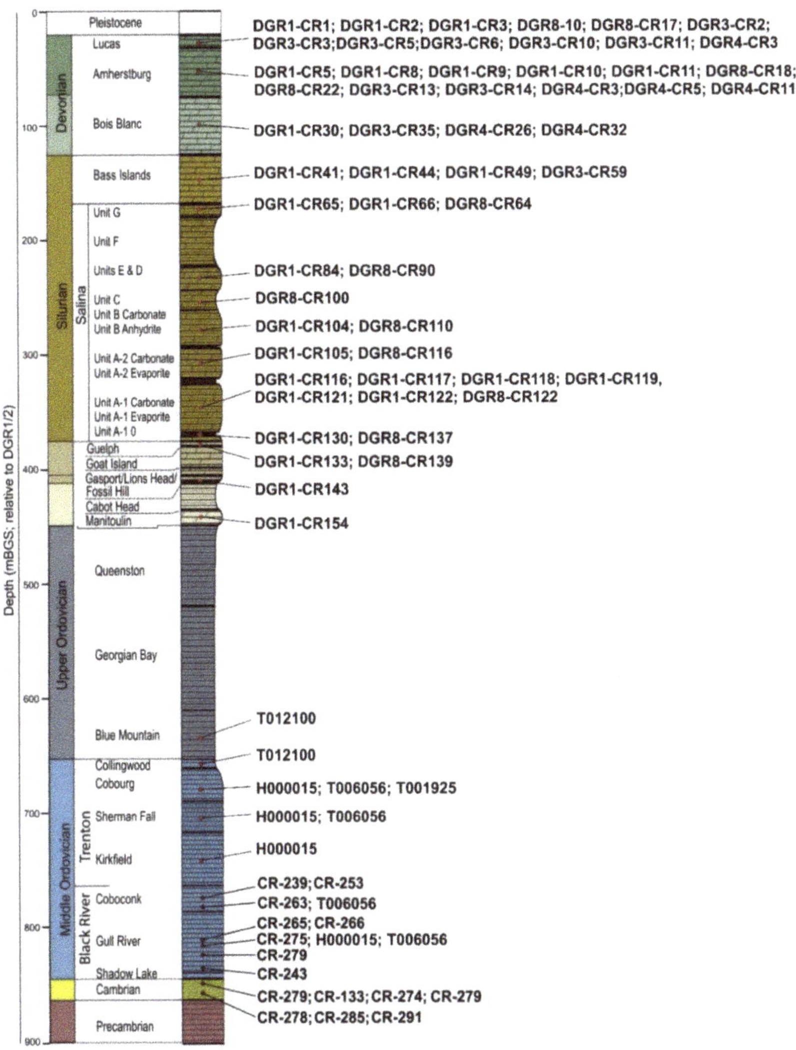

Figure 3. Paleozoic stratigraphy of the studied successions also showing the listing of the studied cores (modified from AECOM [47]).

The Devonian carbonates contain nodular and bioturbated lithofacies, partially dolomitized (e.g., Lucas Formation; Figure 3), algal laminated dolomudstone, dolomitized bioclastic grainstone, undolomitized bioclastic grainstone, and brecciated, cherty dolostone lithofacies.

The diagenetic history of the Cambrian, Ordovician, Silurian, and Devonian successions in the Huron Domain encompasses multiple episodes of calcite cementation, dolomitization, anhydrite cementation, mechanical and chemical compaction, dissolution, and silicification. It is understood (e.g., [2,3]) that these diagenetic processes occurred during deposition and continued upon burial. The paragenetic sequence shown in Figure 4 is based on petrographic relationships, spatial distribution, and geochemical evidence. Due to the significance of dolomitization in these carbonates, the discussion in this paper will be concentrated on the most significant and relevant diagenetic events.

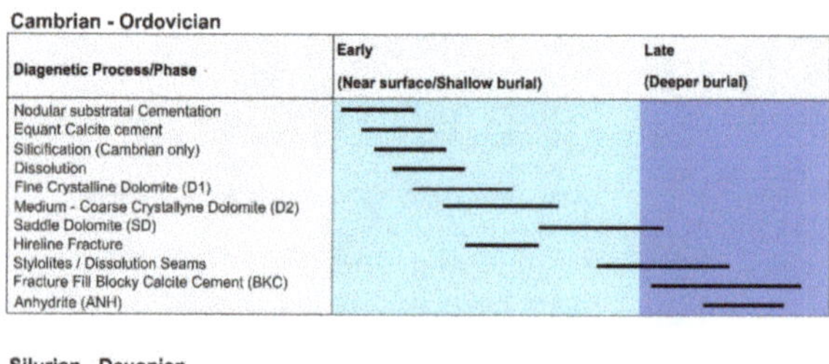

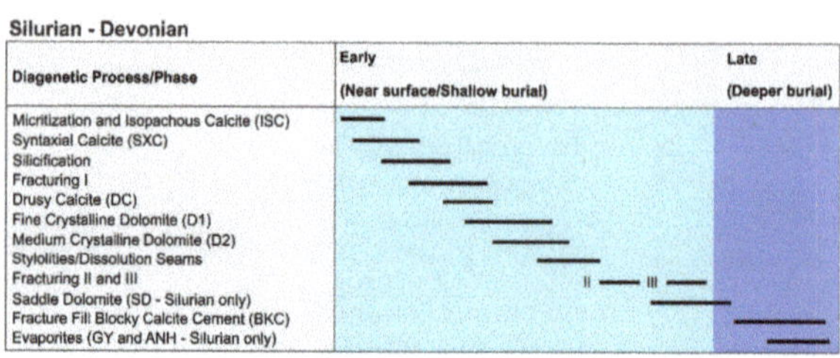

Figure 4. Paragenesis of the Cambrian-Ordovician formations (modified from Al-Aasm and Crowe [2]) and paragenesis of the Silurian-Devonian formations (modified from Tortola et al. [3]).

4.1.1. Dolomite Types

In this study, multiple generations of dolomite have been observed (Figure 4).

The dolomite types occur as both replacements (D) and cement (SD) (cf. [2,3]). According to Sibley and Gregg's classification [68] and based on petrographic characteristics (e.g., crystal size, shape, extinction, fabric, and paragenetic relationships), several types of dolomite development are distinguished. Table 1 summarizes the main types of dolomite and their petrographic and CL characteristics.

Table 1. Summary of the main petrographic characteristics of the main diagenetic minerals in the Paleozoic successions.

Age	Mineral	Description	Texture	Size	Luminescence
Cambrian	D1	Non-ferroan replacive microcrystalline dolomite matrix	Planar-s	<50 µm	Dull red
Cambrian	D2	Non-ferroan medium crystalline dolomite matrix	Planar-s	>50 µm up to 500 µm	Dull red/bright
Cambrian	SD	Minor coarse crystalline saddle dolomite cement filling vugs	Curved crystal faces and undulose extinction	>500 µm	Dull/non-luminescent
Cambrian	BKC	Non ferroan/ferroan calcite cement mainly filling fractures and voids	Blocky/equant	200 µm to >500 µm	Red/bright orange
Ordovician	D1	Non-ferroan replacive microcrystalline dolomite matrix	Non-planar to planar-s	25–50 µm	Dull red
Ordovician	D2	Non-ferroan medium crystalline, partly zoned dolomite matrix	Planar e/planar-s	>50 µm up to 500 µm	Dull red
Ordovician	SD	Minor coarse crystalline saddle dolomite cement filling vugs	Curved crystal faces and undulose extinction	>500 µm	Dull/non-luminescent
Ordovician	SXC	Ferroan syntaxial calcite overgrowth cement around echinoderm fragments		100 µm to >500 µm	Bright red

Table 1. *Cont.*

Age	Mineral	Description	Texture	Size	Luminescence
Ordovician	BKC	Non ferroan/ferroan calcite cement mainly filling fractures and voids	Blocky/equant	200 µm to >500 µm	Red/bright Orange, zoned
Silurian	D1	Non-ferroan pervasive replacive micro to fine crystalline dolomite matrix	Non-planar to planar-s	<50 µm	Dull red/bright
Silurian	D2	Non-ferroan pervasive replacive medium to coarse crystalline dolomite matrix	Planar-e/planar-s	>50 µm up to 150 µm	Dull red/non-luminescent
Silurian	D2 (Dissolution seams)	Selective replacive medium crystalline dolomite matrix (commonly associated with dissolution seams)	Planar-e	>50 µm up to 150 µm	Dull red
Silurian	SD	Ferroan coarse crystalline saddle dolomite cement filling fractures and vugs	Curved crystal faces and undulose extinction	>500 µm	Dull/non-luminescent
Silurian	ISC	Non-ferroan isopachous calcite cement rimming coated grains	bladed	50–100 µm	Dull red
Silurian	SXC	Ferroan syntaxial calcite overgrowth cement around echinoderm fragments	-	100 µm to >500 µm	Bright red
Silurian	DC	Ferroan void-filling and pore-lining cement in intergranular and intraskeletal pores, molds and fractures	Equant to elongate, anhedral to subhedral	75–250 µm Size increase towards the center	Dull red
Silurian	BKC	Non ferroan/ferroan calcite cement mainly filling fractures and voids	blocky	200 µm to >500 µm	Red/bright orange zoned
Devonian	D1	Non-ferroan pervasive replacive micro to fine crystalline matrix dolomite	Non-planar to planar-s	<50 µm	Dull/red
Devonian	D2	Non-ferroan pervasive replacive medium crystalline matrix dolomite	Planar-e/planar-s	>50 µm up to 100 µm	Dull/non-luminescent
Devonian	D2 (Dissolution seams)	Non-feroan selective replacive medium crystalline dolomite matrix (commonly associated with dissolution seams)	Planar-e	>50 µm up to 100 µm	Red/bright
Devonian	SXC	Non-ferroan syntaxial calcite overgrowth cement	-	100 µm to >500 µm	Non-luminescent
Devonian	DTC	Non-ferroan dogtooth calcite cement growing normal to the substrate (mainly skeletal grains)	Elongate scalenohedral or rhombohedral	50 µm	Red/bright mostly zoned
Devonian	DC	Non-ferroan void-filling and pore-lining cement in intergranular and intraskeletal pores, molds and fractures	Equant to elongate, anhedral to subhedral	75–250 µm Size increase towards the center	Dull red to bright, zoned in most cases
Devonian	BKC	Non-ferroan calcite cement mainly filling fractures and voids	blocky	200 µm to >500 µm	Red/bright zoned

Cambrian:

In the mixed carbonate-siliciclastic lithofacies of the Cambrian strata, the presence of dolomite is restricted to the dolomitized bioclastic lithofacies and scattered dolomite crystals within the silicified lithofacies (Figure 5). Two replacive dolomite types occur in the Cambrian rocks (microcrystalline subhedral-D1 and medium to coarse crystalline subhedral to anhedral, zoned- D2; Figure 5) and minor amounts of saddle dolomite (SD) cement present in vugs (Table 1; Figure 5). This dolomite cement is succeeded by large blocky calcite cement (Figure 5I) in fractures and vugs. D2 mimically replaces allochems, such as fossils and ooids.

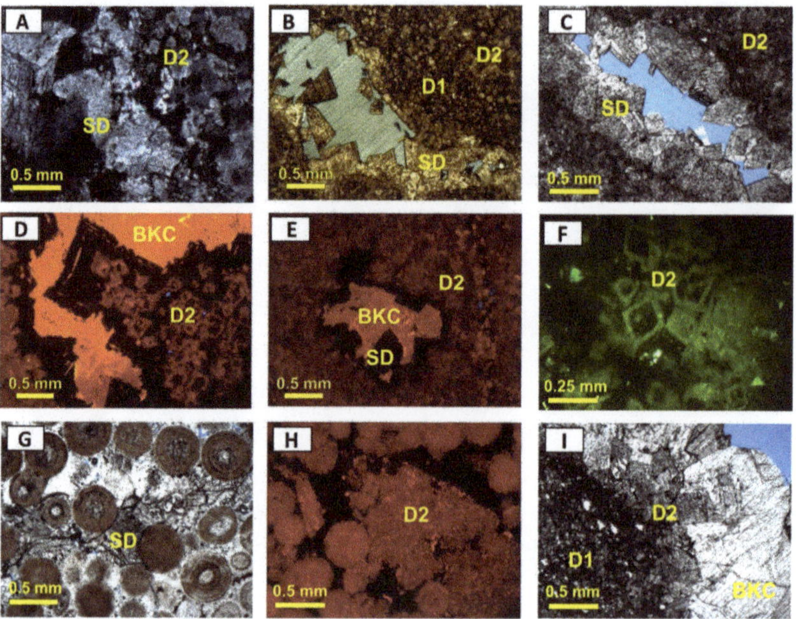

Figure 5. (**A**) Photomicrograph (XPL) of fracture-filling coarse saddle dolomite cement (SD) postdating anhedral, replacive medium to coarse crystalline dolomite matrix (D2). Sample: DGR2-CR279-840.20 Cambrian. (**B**) Photomicrograph (PPL) of fracture-filling coarse saddle dolomite cement (SD) postdating planar-s, pervasive replacive medium to coarse crystalline (D2) and fine crystalline (D1) dolomite matrix. Sample: DGR2-CR139-2-844.87 Cambrian. (**C**) Photomicrograph (XPL) of fracture-filling coarse saddle dolomite cement (SD) postdating anhedral, replacive medium to coarse crystalline dolomite matrix (D2). Sample: DGR2-CR139-2-844.87 Cambrian. (**D**) Photomicrograph (CL) of fracture-filling, bright luminescent blocky calcite cement (BKC) postdating non-luminescent to red luminescent, zoned medium to coarse crystalline dolomite matrix (D2). Sample: DGR4-CR274-2-844.39 Cambrian. (**E**) Photomicrograph (CL) of vug-filling, bright luminescent blocky calcite cement (BKC) postdating non-luminescent saddle dolomite cement (SD) and replacive medium crystalline dolomite matrix (D2) showing dull to red luminescence. Sample: DGR2-CR139-2-844.87 Cambrian. (**F**) Photomicrograph (UV-light) of planar-e pervasive replacive medium to coarse crystalline dolomite matrix (D2) showing dull fluorescent cores and bright rims. Sample: DGR4-CR274-2-844.39 Cambrian. (**G**) Photomicrograph (PPL) of fracture-filling coarse saddle dolomite cement (SD) in an oolitic sandstone. Sample: DGR2-CR134-1-844.54 Cambrian. (**H**) Photomicrograph (CL) of medium to coarse crystalline dolomite matrix (D2) partially replacing coated grains in an oolitic sandstone. Sample: DGR2-CR134-2-844.54 Cambrian. (**I**) Photomicrograph (PPL) of fracture-filling coarse blocky calcite cement (BKC) postdating planar-s, pervasive replacive medium to coarse crystalline (D2) and fine crystalline (D1) dolomite matrix. Sample: DGR2-CR133-1-843.96 Cambrian.

Ordovician:

Two types of replacive dolomite are recognized in the Gull River, Coboconk, and Cobourg formations (Table 1; Figure 6): (1) microcrystalline dolomite replacing skeletal components and matrix (D1), and (2) medium crystalline replacive matrix dolomite (D2), which is fabric destructive, partly zoned and replaces the intranodular micritic matrix (Figure 6). D2 dolomite is volumetrically more abundant than D1 and it is also observed along dissolution seams in partially dolomitized limestones, usually characterized by cloudy-dark cores and clear rims (Figure 6C). A minor presence of saddle dolomite (SD) occludes pores and fractures and postdates D1 and D2 (Figure 6D,E). Anhydrite cement postdates saddle dolomite (Figure 6E,F). In other Ordovician formations, such as the Shadow Lake, Kirkfield and Sherman Fall, Silurian, which are mostly dominated by limestones of varied lithologies, no significant dolomitization is observed. However, the scattered occurrence of D1 dolomite is observed in Sherman Fall Formation replacing the calcitic muddy matrix.

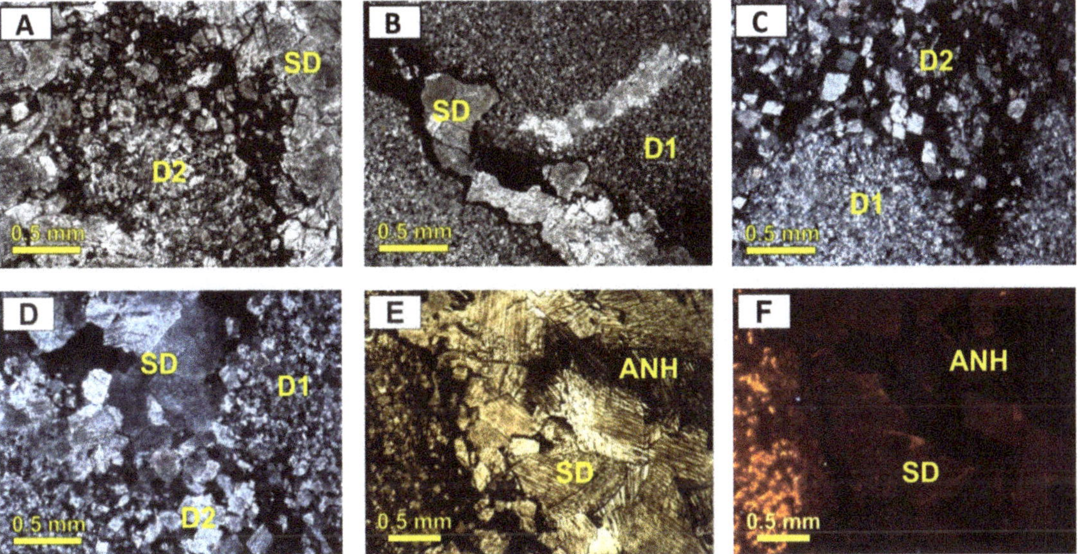

Figure 6. (**A**) Photomicrograph (PPL) of fracture-filling coarse saddle dolomite cement (SD) postdating planar-e to planar-s, replacive medium to coarse crystalline(D2) dolomite matrix. Sample: DGR3-CR279-2-839.80 Gull River. (**B**) Photomicrograph (PPL) of fracture-filling coarse saddle dolomite cement (SD) postdating, pervasive replacive micro to fine crystalline dolomite matrix (D1). Sample: DGR3-CR279-2-839.80 Gull River. (**C**) Photomicrograph (XPL) of planar-e replacive medium to coarse crystalline dolomite matrix (D2) and fine crystalline dolomite matrix (D1). Sample: T001925-1-890 (1-5) Cobourg. (**D**) Photomicrograph (XPL) of fracture-filling coarse saddle dolomite cement (SD) postdating planar-s, pervasive, replacive medium to coarse (D2) and fine crystalline (D1) dolomite matrix. Sample: DGR4-CR265-817.12 Gull River. (**E**,**F**) Paired photomicrographs (PPL and CL, respectively) of anhydrite cement postdating zoned, dull (rims) to dark red (cores) luminescent, coarse saddle dolomite cement (SD). Sample: DGR3-CR279-2-839.80 Gull River.

Silurian:

Dolomite in the Silurian succession occurs mostly in the Guelph, Salina (A1-carbonates, A2-carbonates, Figure 3), and Bass Island formations. Two types of replacive matrix dolomite and one of dolomite cement are observed (Table 1; Figure 7). These are: D1 which is a pervasive replacive micro to fine crystalline matrix dolomite and D2, which consists of medium crystalline, subhedral, zoned dolomite (Figure 7B,C). It also occurs as a less abundant selective replacive medium (>50–100 μm) euhedral crystalline dolomite, only observed along dissolution seams in partially dolomitized limestones and usually characterized by cloudy-dark cores and clear rims (Figure 7C). Ferroan saddle dolomite (SD; Figure 7D–F) occludes fractures and vugs predating late blocky calcite cement (Figure 7E,F).

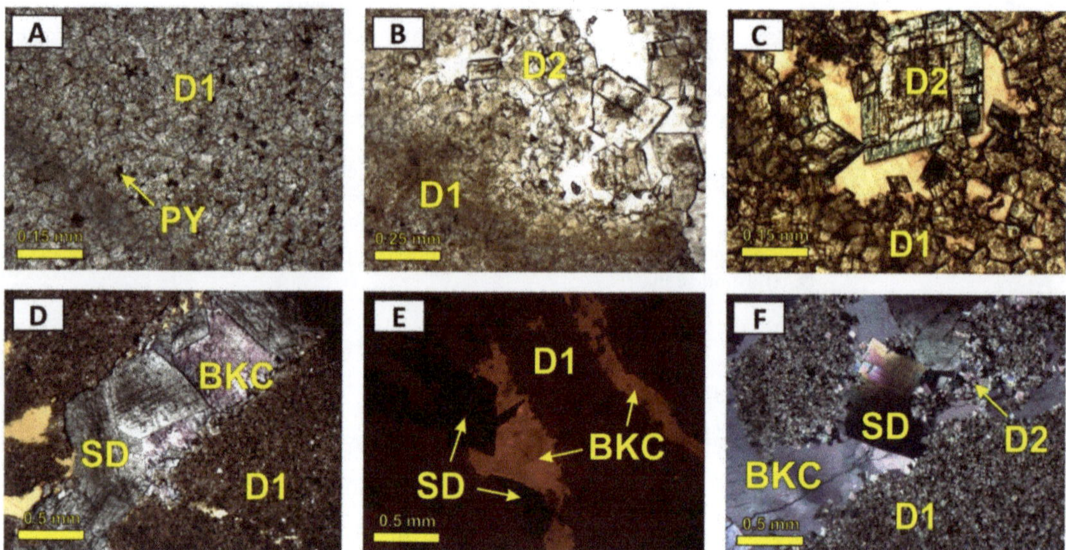

Figure 7. (**A**) Photomicrograph (PPL) of planar-s replacive micro to fine crystalline dolomite matrix (D1). The yellow arrow indicates the presence of framboidal pyrite in the matrix. Sample: DGR3-CR61-194.02 (3-11) Salina G-Unit. (**B**) Photomicrograph (PPL) of planar-s fine crystalline dolomite matrix (D1) and planar-e to planar-s, replacive medium crystalline dolomite matrix (D2), showing increasing intercrystalline porosity. Sample: DGR4-CR100-327.98 (4-11) Salina A1-Unit (Carbonate). (**C**) Photomicrograph (PPL—after staining) of planar-s fine crystalline dolomite and planar-e to planar-s, pervasive replacive medium crystalline dolomite matrix (D2) showing increasing intercrystalline porosity and non-ferroan cores followed by later iron-rich rims. Sample: DGR3-CR128-388.45 (3-16) Guelph. (**D**) Photomicrograph (PPL—after staining) of fracture-filling blocky calcite cement (BKC) postdating fracture-filling coarse saddle dolomite cement (SD) and fine crystalline (D1) dolomite matrix. Sample: DGR8-CR139-382.40 (8-13) Guelph. (**E**) Photomicrograph (CL) of fracture-filling, bright luminescent blocky calcite cement (BKC) postdating dull to non-luminescent coarse saddle dolomite cement (SD) and dark red luminescent fine crystalline dolomite matrix (D1). Sample: DGR8-CR139-382.40 (8-13) Guelph. (**F**) Photomicrograph (XPL) of fracture-filling blocky calcite cement (BKC) postdating fracture-filling coarse saddle dolomite cement (SD), replacive medium to coarse crystalline dolomite matrix (D2) and fine crystalline dolomite matrix (D1). Sample: DGR8-CR116-313.31 (8-9) Salina A2-Unit (Carbonate).

Devonian:

Two types of dolomites occur in the Devonian formations (Figure 3). These are: a pervasive replacive euhedral to subhedral micro to fine crystalline matrix dolomite (D1) and a pervasive replacive euhedral to subhedral medium crystalline, zoned matrix dolomite (D2) that is also associated sometimes associated with dissolution seams (Figure 8E). They are primarily observed in the Lucas and Amherstburg formations (Table 1; Figure 8).

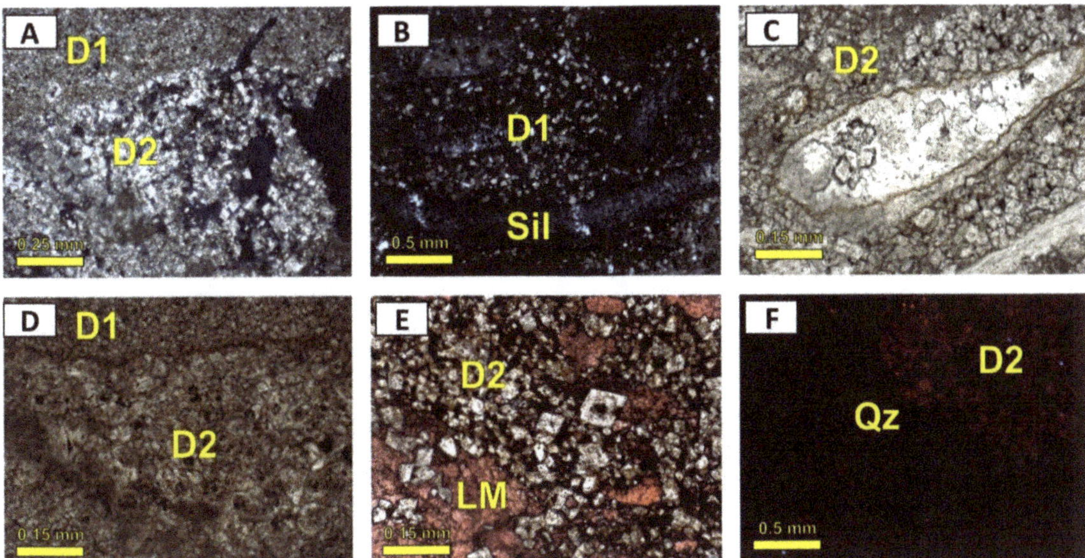

Figure 8. (**A**) Photomicrograph (XPL) of planar-s fine crystalline dolomite matrix (D1) and planar-e to planar-s, replacive medium crystalline dolomite matrix (D2), showing increasing intercrystalline porosity. DGR8-CR22-59.06 (8-4) Amherstburg. (**B**) Photomicrograph (XPL) of chert (Sil) and silicified skeletal fragment partially replaced by fine crystalline dolomite (D1). Sample: DGR4-CR26-105.11 (4-5) Bois Blanc. (**C**) Photomicrograph (PPL) of planar-e to planar-s, medium crystalline dolomite matrix (RD2) pervasively replacing precursor limestone and partially replacing allochems and skeletal fragments. Sample DGR1-CR2-3-27.75 (1-4) Lucas. (**D**) Photomicrograph (PPL) of micro to fine crystalline dolomite matrix (D1) and planar-e to planar-s, replacive medium crystalline dolomite matrix (D2) associated with dissolution seams showing dark cores and clear rims. Sample: DGR3-CR6-42.30 (3-4) Lucas. (**E**) Photomicrograph (PPL—after staining) of planar-e, replacive medium crystalline dolomite matrix (D2) associated with dissolution seams Characterized by cloudy cores and clear rims. The undolomitized limestone matrix (LM) shows a pink color after staining. Sample: DGR1-CR3-2-31.85 (1-6) Amherstburg. (**F**) Photomicrograph (CL) of fracture-filling authigenic quartz (Qz) postdating dull to dark red luminescent medium crystalline dolomite matrix (D2). Sample: DGR4-CR26-105.11 (4-5) Bois Blanc.

4.1.2. Calcite Cementation

Early and late calcite type of cement are present in the investigated Paleozoic formations. These types of cement display various crystal habits and sizes and occlude both primary as well as secondary porosity, pre- and post-dolomitization (Figures 4 and 9). The types of calcite cement identified include isopachous, syntaxial overgrowth, equant, dog-tooth, and blocky cement (Table 1).

Within Cambrian strata, fractures are occluded by a coarse, equant/ blocky calcite cement with strong, bright CL (Figure 9). In contrast, calcite cementation in the Ordovician formations is represented by an early equant calcite cement surrounding fossil allochems, isopachous rims around fossils and occluding hairline fractures (Figure 9), and later blocky calcite in fractures and vugs. The blocky calcite (BKC) in fractures postdates saddle dolomite cement (Figure 9).

In the Silurian successions, isopachous calcite cement growing around coated grains and exhibits microcrystalline, bladed texture (Figure 9; Table 1). Syntaxial calcite overgrowth cement (SXC; Figure 9D) forms ferroan, bright-luminescent crystals. Void-filling equant, pore-lining drusy calcite cement is observed in intergranular and intraskeletal pores, molds, and fractures. Late fracture and void-filling blocky calcite cement (BKC) postdate saddle dolomite cement. In terms of paragenesis, early calcite cement predated D1 and D2 and predated fracture- and pore-filling saddle dolomite which is postdated by late blocky calcite (BKC; Figures 4, 7F and 9) and evaporite cement.

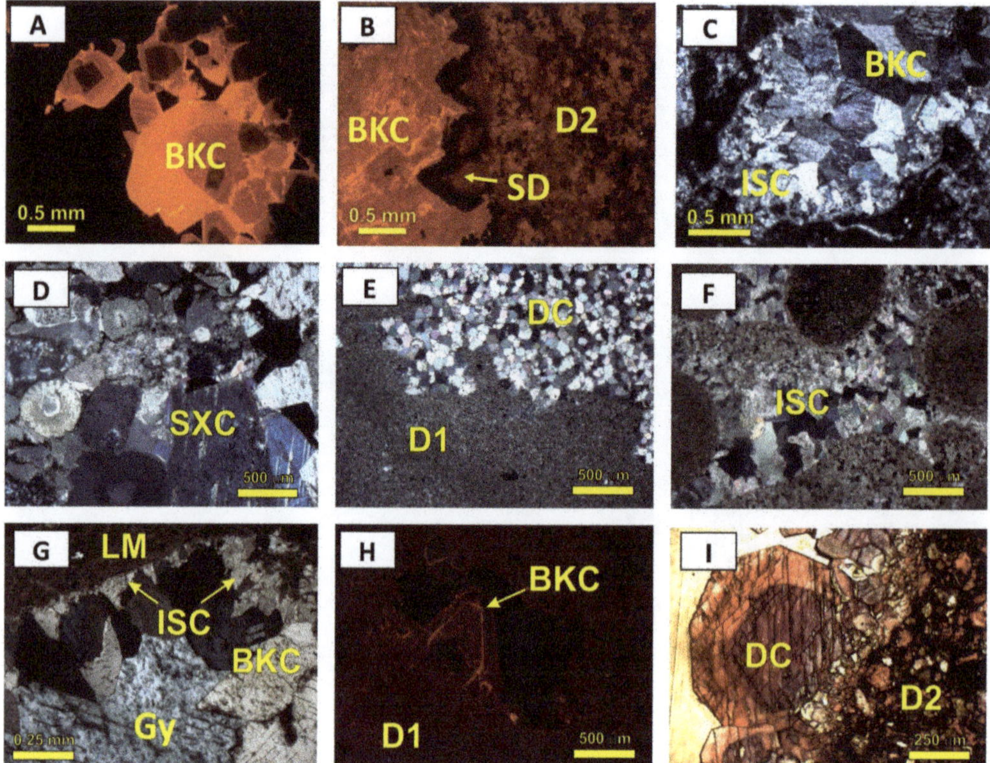

Figure 9. (**A**) Photomicrograph (CL) of fracture-filling, zoned, blocky calcite cement (BKC) characterized by dull to non-luminescent cores and bright luminescent rims. Sample: T006056- 57-396.1 (2-15) Coboconk. (**B**) Photomicrograph (CL) of fracture-filling, bright luminescent blocky calcite cement (BKC) postdating zone, dull to non-luminescent coarse saddle dolomite cement (SD) and dark red luminescent medium crystalline dolomite matrix (D2). Sample: DGR2-CR133-1-843.96 Cambrian. (**C**) photomicrograph (XPL) of bladed vug-lining isopachous calcite cement (ISC) predating vug-filling blocky calcite cement (BKC). Sample: DGR6-CR243-898.24 Gull River. (**D**) Photomicrograph (XPL) of syntaxial overgrowth calcite cement (SXC) on crinoidal and echinoid skeletal fragments in a bioclastic grainstone. Sample: DGR1-CR154-440.92 (1-33) Manitoulin. (**E**) Photomicrograph (XPL) of drusy calcite cement (DC) and micro to fine crystalline dolomite matrix (D1) in fenestral dolostone. Sample: DGR1-CR49-142.28 (1-17) Bass Island. (**F**) Photomicrograph (XPL) of bladed isopachous calcite cement (ISC) filling interparticle pores between coated grains in a partially dolomitized grainstone. DGR8-CR122-332.26 (8-10) Salina A1-Unit (Carbonate). (**G**) Photomicrograph (XPL) of bladed vug-lining isopachous calcite cement (ISC) predating vug-filling blocky calcite cement (BKC) and gypsum (Gy). Sample: DGR1-CR119-337.32 (1-27) Salina A1-Unit (Carbonate). (**H**) Photomicrograph (CL) of fracture-filling, zoned, blocky calcite cement (BKC) postdating dark red luminescent fine crystalline dolomite matrix (D1). Sample: DGR3-CR10-51.60 (3-5) Lucas. (**I**) Photomicrograph (PPL—after staining) of non-ferroan, drusy calcite cement (DC) lining a fine crystalline dolostone. The coarser crystals show zonation characterized by iron-rich cores followed by non-ferroan rims. Sample: DGR3-CR335.35 (3-2) Lucas.

In the Devonian formations, four types of calcite cement occlude the pore spaces represented by syntaxial overgrowth, dog-tooth, equant/drusy, and blocky. Paragenetically, early calcite cement (e.g., syntaxial overgrowth, dog-tooth; Figures 4 and 9) predate replacive dolomite matrix D1 and predate the late fracture- and pore-filling blocky calcite (BKC) cement.

4.1.3. Other Diagenetic Phases

Together with the calcite and dolomite diagenetic phases reported above, silica-bearing minerals and sulfate minerals are also present in the studied successions (Figure 10).

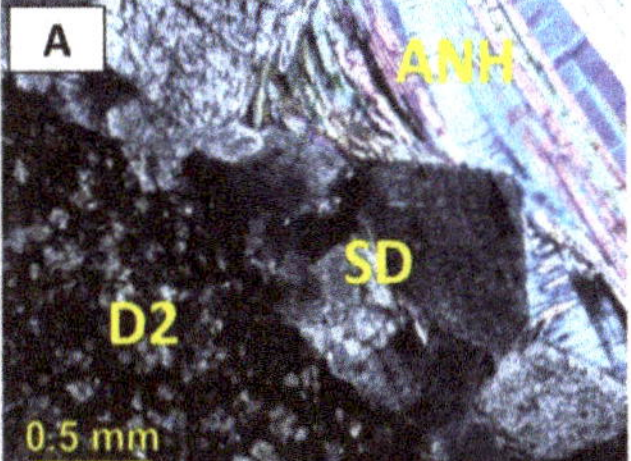

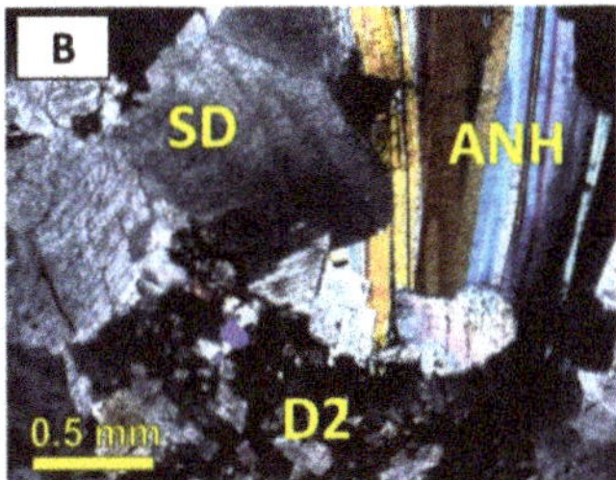

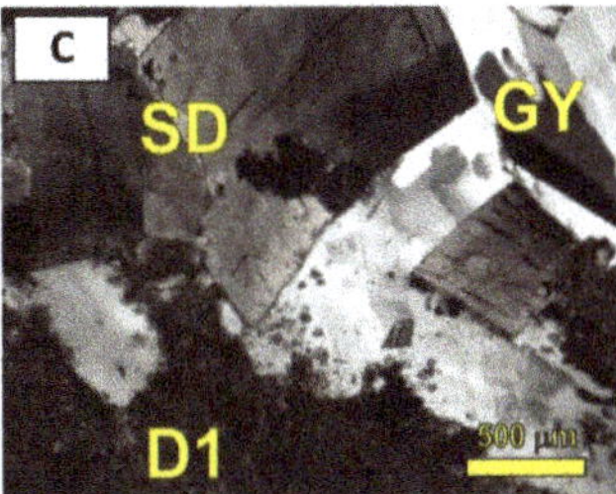

Figure 10. (**A**) Photomicrograph (XPL) of anhydrite cement (ANH) postdating coarse saddle dolomite cement (SD) and planar-s medium crystalline dolomite matrix (D2). Sample: DGR3-CR279-840.20 Cambrian. (**B**) Photomicrograph (XPL) of anhydrite cement (ANH) postdating coarse saddle dolomite cement (SD) and planar-e to planar-s medium crystalline dolomite matrix (D2). Sample: DGR3-CR279-2-839.80 Gull River. (**C**) Photomicrograph (XPL) of gypsum cement (Gy) postdating coarse saddle dolomite cement (SD) and micro to fine crystalline dolomite matrix (D1). Sample: DGR8-CR116-313.31 (8-9) Salina A2-Unit (Carbonate).

Quartz is common in the Cambrian rocks typically replacing skeletal and nonskeletal components such as corals in coral-bearing facies and coated grains, such as ooids (Figure 5G). Silicification pre-dates dolomitization but postdates early isopachous calcite cementations (Figure 4). In the Ordovician strata, silicification replaced part of the muddy calcitic matrix, with microcrystalline dolomite (D1) replacing the silicified micrite and fossils. Anhydrite and gypsum are present in Cambrian, Ordovician, and Silurian rocks as fracture-fill, which postdates replacive dolomite (Figure 10). The rare occurrence of authigenic K-feldspar is observed by SEM filling some pores in Ordovician and Silurian rock but predated D2. Also, very minor pyrite and fluorite are observed in Silurian dolostones.

4.2. Geochemical and Fluid Inclusion Results

4.2.1. Oxygen, Carbon and Sr Isotopes

Table 2 summarizes the stable and radiogenic isotope results of dolomite and calcite cement types. Figures 11–15 show the relationships among these isotopes with respect to their petrographic characteristics.

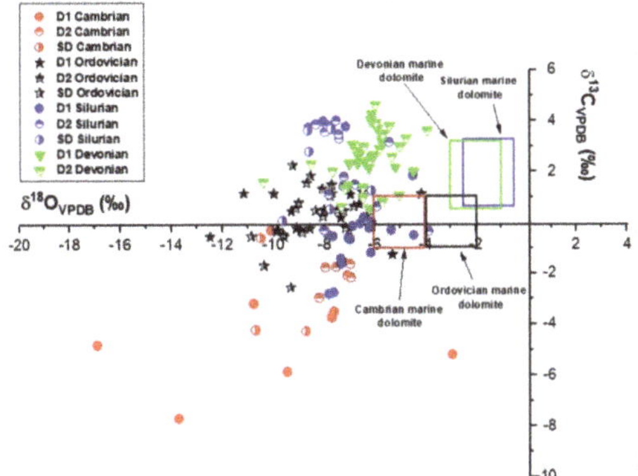

Figure 11. Cross plot of $\delta^{13}C_{VPDB}$ vs. $\delta^{18}O_{VPDB}$ of different generations of dolomite. Note the overlap in many values and the shift from the postulated range for respective seawater composition.

Table 2. Summary of the isotopic composition of dolomite and calcite components in the studied successions.

Age	Phase	$\delta^{13}C_{VPDB}$ (‰)	$\delta^{18}O_{VPDB}$ (‰)	$^{87}Sr/^{86}Sr$
Cambrian	D1			
	n.	8	8	5
	Avg.	−4.27	−9.91	0.71128
	Stdev.	2.03	3.93	0.00298
	Max.	−0.30	−2.92	0.71716
	Min.	−7.67	−16.91	0.70928
Cambrian	D2			
	n.	7	7	4
	Avg.	−1.98	−7.42	0.70998
	Stdev.	0.45	0.46	0.00068
	Max.	−1.56	−6.95	0.71100
	Min.	−2.96	−8.20	0.70927
Cambrian	SD			
	n.	3	3	0
	Avg.	−3.02	−9.98	
	Stdev.	1.72	0.89	
	Max.	−0.59	−8.72	
	Min.	−4.26	−10.71	
Cambrian	BKC			
	n.	4	4	3
	Avg.	−4.47	−12.27	0.71011
	Stdev.	0.63	2.85	0.00014
	Max.	−3.55	−7.93	0.71031
	Min.	−5.30	−15.20	0.71000
Ordovician	D1			
	n.	15	15	7
	Avg.	0.11	−8.87	0.71189
	Stdev.	0.70	1.77	0.00517
	Max.	1.16	−5.30	0.72346
	Min.	−1.25	−12.48	0.70808
Ordovician	D2			
	n.	19	19	7
	Avg.	0.70	−7.71	0.70896
	Stdev.	0.71	1.26	0.00148
	Max.	2.24	−4.15	0.71255
	Min.	−0.46	−9.62	0.70814
Ordovician	SD			
	n.	6	6	2
	Avg.	−0.04	−9.44	0.70955
	Stdev.	1.75	0.91	0.00111
	Max.	2.24	−8.33	0.71066
	Min.	−2.56	−10.86	0.70844
Ordovician	Early Calcite			
	n.	41	41	1
	Avg.	0.30	−5.67	
	Stdev.	0.85	1.24	
	Max.	2.16	−3.89	0.70809
	Min.	−3.19	−9.21	0.70809
Ordovician	BKC			
	n.	30	30	4
	Avg.	−0.15	−6.78	0.70801
	Stdev.	0.70	1.20	0.00005
	Max.	1.46	−3.94	0.70809
	Min.	−1.84	−9.88	0.70797

Table 2. *Cont.*

Age	Phase	$\delta^{13}C$ VPDB (‰)	$\delta^{18}O$ VPDB (‰)	$^{87}Sr/^{86}Sr$
Silurian	D1			
	n.	21	21	8
	Avg.	0.21	−6.83	0.70935
	Stdev.	1.99	1.03	0.00103
	Max.	3.99	−4.46	0.71201
	Min.	−2.80	−8.03	0.70871
Silurian	D2			
	n.	14	14	7
	Avg.	2.23	−7.06	0.70867
	Stdev.	1.51	1.18	0.00013
	Max.	4.02	−3.85	0.70891
	Min.	−0.32	−8.31	0.70847
Silurian	SD			
	n.	6	6	3
	Avg.	2.00	−8.08	0.70863
	Stdev.	1.44	1.08	0.00008
	Max.	3.69	−6.13	0.70872
	Min.	0.10	−9.65	0.70852
Silurian	Early Calcite			
	n.	12	12	1
	Avg.	0.96	−6.65	
	Stdev.	1.21	1.38	
	Max.	3.53	−4.73	0.70815
	Min.	−0.88	−8.85	0.70815
Silurian	BKC			
	n.	10	10	3
	Avg.	−0.26	−7.92	0.70820
	Stdev.	1.98	1.66	0.00014
	Max.	3.17	−4.54	0.70838
	Min.	−3.75	−10.19	0.70805
Devonian	D1			
	n.	16	16	8
	Avg.	2.65	−6.16	0.70832
	Stdev.	0.90	0.66	0.00059
	Max.	4.28	−4.46	0.70937
	Min.	0.55	−7.04	0.70788
Devonian	D2			
	n.	22	22	8
	Avg.	2.50	−6.22	0.70873
	Stdev.	1.11	1.35	0.00053
	Max.	4.63	−3.91	0.70937
	Min.	0.64	−10.39	0.70809
Devonian	Early Calcite			
	n.	19	19	1
	Avg.	1.45	−6.33	
	Stdev.	1.48	1.21	
	Max.	4.12	−4.42	0.70805
	Min.	−2.42	−9.69	0.70805
Devonian	BKC			
	n.	5	5	4
	Avg.	−1.58	−9.12	0.70803
	Stdev.	0.43	0.85	0.00003
	Max.	−1.22	−8.01	0.70806
	Min.	−2.40	−10.54	0.70798

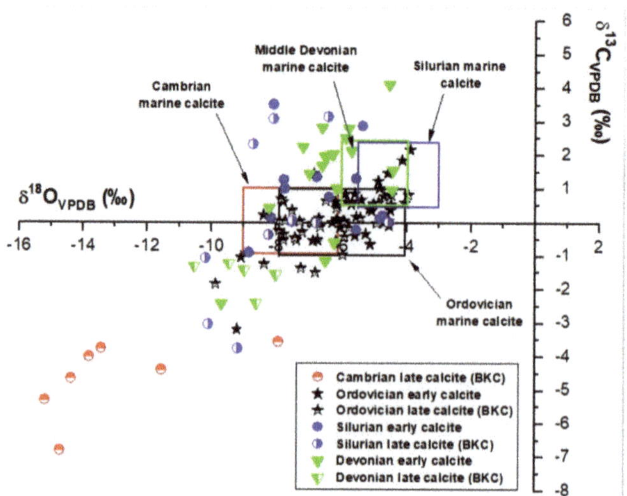

Figure 12. Cross plot of $\delta^{13}C_{VPDB}$ vs. $\delta^{18}O_{VPDB}$ of different generations of calcite cement of different ages. Note the overlap in many values and the negative shift of late calcite from the postulated range for respective seawater composition.

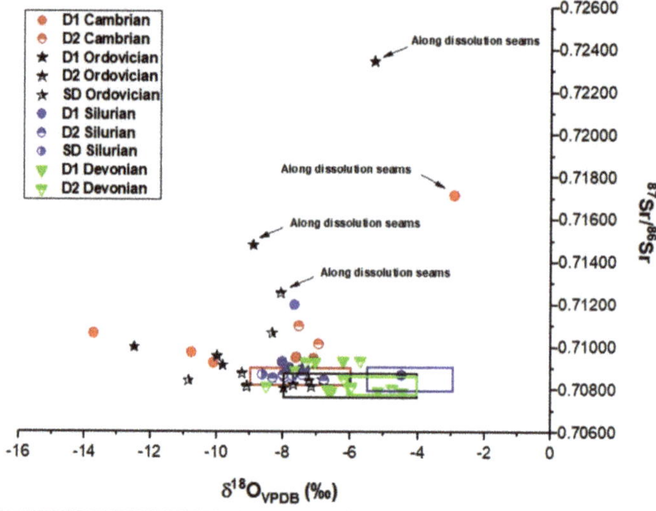

Figure 13. Cross plot of $\delta^{18}O_{VPDB}$ values vs. $^{87}Sr/^{86}Sr$ ratios of different generations of dolomite. Note the overlap in many values and the shift from the postulated values for respective seawater composition except for dolomite associated with dissolution seams.

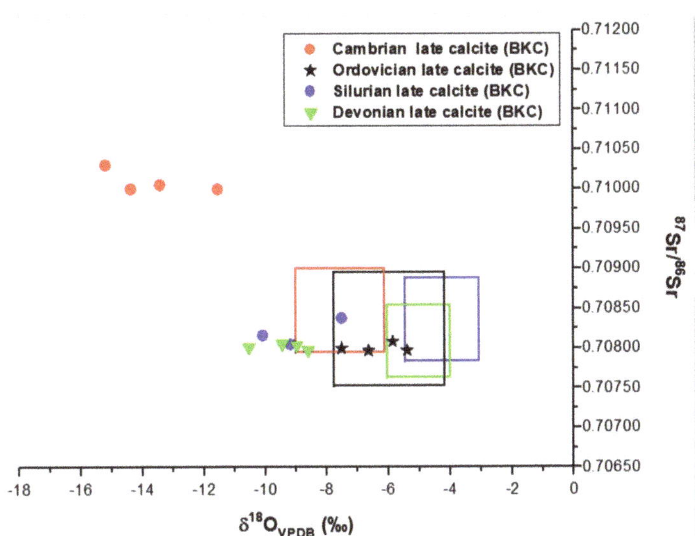

Figure 14. Cross plot of $\delta^{18}O_{VPDB}$ values vs. $^{87}Sr/^{86}Sr$ ratios of different generations of calcite cement. Note the highly radiogenic ratios in Cambrian blocky calcite.

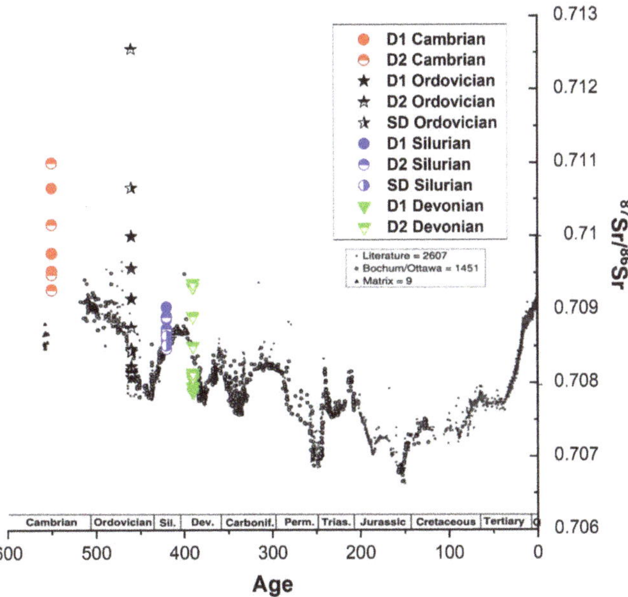

Figure 15. Comparison between $^{87}Sr/^{86}Sr$ secular variation [69] and $^{87}Sr/^{86}Sr$ ratios of the investigated dolomite.

Fine-crystalline matrix dolomite in the Cambrian carbonates (D1) has a wide range of $\delta^{18}O$ values from −16.9 to −2.9‰, $\delta^{13}C$ values from −7.6 to −0.3‰ and $^{87}Sr/^{86}Sr$ ratios from 0.70928 to 0.71716, respectively. Medium crystalline dolomite (D2) $\delta^{18}O$ values vary from −8.2 to −6.9 ‰, $\delta^{13}C$ values vary from −2.9 to −1.5‰, and $^{87}Sr/^{86}Sr$ ratios from 0.70927 to 0.71100. Saddle dolomite cement $\delta^{18}O$ values range between −10.7 to −8.7‰, from −4.2 to −0.5‰ for $\delta^{13}C$, respectively. In all dolomite types there is a co-variant relationship between both isotopes (Figure 11). Blocky calcite cement (BKC) occluding fractures have $\delta^{18}O$ values ranging from −15.2 to −7.9‰ for $\delta^{18}O$ and $\delta^{13}C$ values −5.3 to −3.5‰. Its $^{87}Sr/^{86}Sr$ ratios range from 0.71000 to 0.71031 (Figure 12).

The fine crystalline replacive matrix dolomite (D1) in the Ordovician carbonates has $\delta^{18}O$ values vary from −12.4 to −5.3‰, and less negative $\delta^{13}C$ values compared to the Cambrian D1 dolomite (ranging from −1.2 to 1.1‰; Figure 11). Its $^{87}Sr/^{86}Sr$ ratios range from 0.70808 to 0.72346, respectively. The highly enriched ratios are samples that occur along dissolution seams (Figure 13). D2 overlaps with D1 with $\delta^{18}O$ values ranging from −9.6 to −4.1‰, $\delta^{13}C$ values from −0.4 to 2.2‰, and $^{87}Sr/^{86}Sr$ ratios from 0.70814 to 0.71255, respectively. The enriched ratio also represents a sample occurring along dissolution seams (Figure 13). The $\delta^{18}O$ values for saddle dolomite (SD) vary from −10.8 to −8.3‰, and $\delta^{13}C$ values from −2.5 to 2.2‰, respectively. Its $^{87}Sr/^{86}Sr$ ratios vary from 0.70844 to 0.71066. The $\delta^{18}O$ values for the early calcite matrix ranging from −9.2 to −3.8‰, and from −3.9 to 2.1‰ for carbon isotopes. One sample $^{87}Sr/^{86}Sr$ ratios for matrix calcite is 0.70809. For late blocky calcite cement (BKC) the $\delta^{18}O$ values vary from −9.8 to −3.9 and from −1.8 to 1.4‰ for $\delta^{13}C$. The $^{87}Sr/^{86}Sr$ ratios for this cement vary from 0.70797 to 0.70809.

There is a significant overlap in $\delta^{18}O$ and $\delta^{13}C$ values in D1, D2, and SD. Fine-crystalline matrix dolomite in the Silurian carbonates (D1) has $\delta^{18}O$ values ranging from −8.0 to −4.4‰, $\delta^{13}C$ values from −2.8 to 3.9‰ and $^{87}Sr/^{86}Sr$ ratios from 0.70871 to 0.71201, respectively. Medium crystalline dolomite (D2) $\delta^{18}O$ values vary from −8.3 to −3.8, $\delta^{13}C$ values vary from −0.3 to 4.0‰, and $^{87}Sr/^{86}Sr$ ratios from 0.708447 to 0.70891. Saddle dolomite cement $\delta^{18}O$ values range between −9.6 to −6.1‰, from 0.1‰ to 3.69 for $\delta^{13}C$, respectively. Its $^{87}Sr/^{86}Sr$ ratios vary from 0.70852 to 0.70872. The $\delta^{18}O$ values for the early calcite matrix ranging from −8.8 to −4.7‰, and from −0.8 to 3.5‰ for carbon isotopes. In one sample, $^{87}Sr/^{86}Sr$ ratios for the matrix calcite is 0.70815. Blocky calcite cement occluding fractures has $\delta^{18}O$ values ranging from −10.1 to −4.5 for $\delta^{18}O$ and $\delta^{13}C$ values −3.75 to 3.1‰. Its $^{87}Sr/^{86}Sr$ ratios range from 0.70805 to 0.70838.

Most of the $\delta^{18}O$ values for D1 in the Devonian carbonates are slightly more enriched than D2 and they vary between −7.0 to −4.4 for $\delta^{18}O$ and from 0.5 to 4.2 for $\delta^{13}C$, respectively. Its $^{87}Sr/^{86}Sr$ ratios range from 0.70788 to 0.70937. Medium crystalline dolomite (D2) $\delta^{18}O$ values vary from −10.3 to −3.9, $\delta^{13}C$ values vary from 0.6 to 4.6‰, and $^{87}Sr/^{86}Sr$ ratios from 0.70809 to 0.70937. The $\delta^{18}O$ values for the early calcite matrix ranging from −9.6 to −4.4‰, and from 2.4 to 4.1‰ for carbon isotopes. The one analysis of Sr isotopes of this calcite yields a ratio of 0.70805. Blocky calcite cement (BKC) occluding fractures and vugs has $\delta^{18}O$ values ranging from −10.5 to −8.0 for $\delta^{18}O$ and $\delta^{13}C$ values −2.4 to −1.2‰. Its $^{87}Sr/^{86}Sr$ ratios range from 0.70798 to 0.70806.

4.2.2. Major, Minor, Trace and REE Elements Content

Analytical results of major, trace, and rare-earth (REE) element concentrations for the dolomites and calcites from various age groups are summarized in Table 3. All analyzed replacement and cement dolomite types are non-stoichiometric, consistently calcium-rich (average $CaCO_3$ ranging from 54.3 to 61.8 mol% for D1 and D2, and from 57.7 to 63.3 mol% for SD, respectively; Figure 16). There are no significant differences in Mg content in zoned D2 and SD between the cores and rims, as evidenced from the semiquantitative EDX analysis. In general, Mn and Fe contents in the Cambrian and Ordovician dolomites are much higher than the Silurian and Devonian dolomites (Table 3). There is also a covariant trend between Mn and Fe for these dolomites (Figure 17). However, the average of Sr content in all dolomites does not show large variations among the investigated dolomite types (vary between 83 and 253 ppm, Table 3), but many samples from the Ordovician dolomites show high Sr content and covariant trends with Fe and Mn (Figure 17). Such a relationship is not apparent in the dolomites from other age groups.

Table 3. Summary of the major and trace element composition of the studied successions.

Age-Phase	Statistics	$CaCO_3$	Mn	Fe	Sr
Cambrian-D1 (n = 4)	Avg.	59.3	1558.3	8912.8	83.6
	Stdev.	4.2	400.4	2013.8	56.2
	Max.	66.5	1945	11,672.6	180.4
	Min.	56	926.9	6773.7	44.8
Cambrian-D2 (n = 6)	Avg.	55.3	907.2	3883.6	97.8
	Stdev.	0.3	163.7	767.4	6.3
	Max.	55.7	1227.7	5364.1	107.7
	Min.	734.3	2942.3	90.6	0
Cambrian-SD (n = 1)	Avg.				
	Stdev.				
	Max.	57.7	1026.4	12,579.1	89.5
	Min.	57.7	1026.4	12,579.1	89.5
Ordovician-D1 (n = 5)	Avg.	58.4	1097.9	8115.6	90.9
	Stdev.	1.2	430.4	1787.9	43.2
	Max.	60.2	1537.4	10,521.4	172
	Min.	56.7	341.8	5462.3	55.5
Ordovician-D2 (n = 13)	Avg.	61.8	793.3	10,128.9	120.9
	Stdev.	5.7	347	8459	59.2
	Max.	75	1305.5	29,800	236.4
	Min.	56.5	409.5	2942.8	41.5
Ordovician-SD (n = 3)	Avg.	63.3	1157.6	15,055.2	108.6
	Stdev.	7.2	227.7	6420.9	38.1
	Max.	73.5	1459.1	23,597.3	154.9
	Min.	57.1	909	8116.3	61.4
Silurian-D1 (n = 16)	Avg.	54.3	223.8	2428	94.6
	Stdev.	3.4	217.9	1729.4	18.7
	Max.	59.4	857.8	7017.6	125.7
	Min.	48.3	34.2	339.3	70.2
Silurian-D2 (n = 7)	Avg.	55.9	433.3	4621.3	253.7
	Stdev.	3.1	313.2	2423.3	396.5
	Max.	62.1	860.9	8249.6	1222.1
	Min.	52.6	66.3	809.7	70.7
Silurian-SD (n = 3)	Avg.	62.3	1098.9	19,606.4	105.9
	Stdev.	4.4	631.1	1132.4	18
	Max.	67	1645.5	20,439.8	124.8
	Min.	56.3	214.6	18,005.3	81.8
Devonian-D1 (n = 13)	Avg.	55.6	33.2	219.8	71.8
	Stdev.	4.4	11.2	239.8	17.4
	Max.	62.2	54.2	690.5	123.5
	Min.	50	14.3	33.9	46.8
Devonian-D2 (n = 9)	Avg.	56.8	48	419.9	78.2
	Stdev.	4.1	10.7	296.6	12.2
	Max.	61.6	60.8	791.4	97.1
	Min.	47.9	32.9	83	60.3

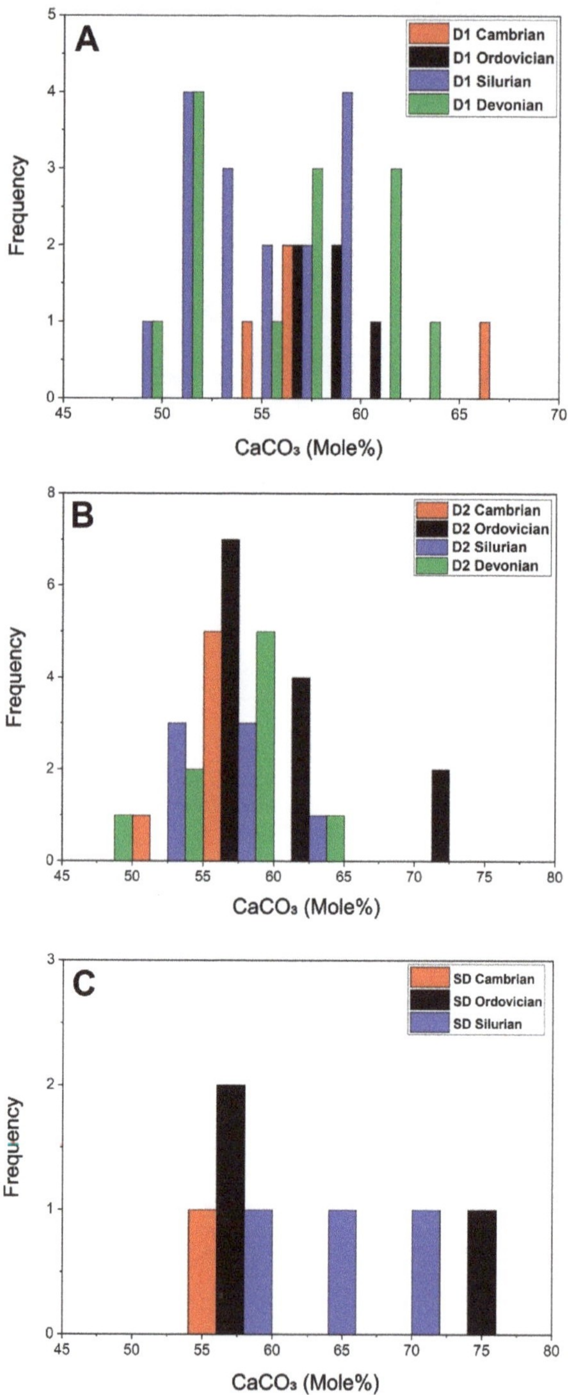

Figure 16. Histogram plots of $CaCO_3$ (mol%) of: (**A**) D1; (**B**) D2, and (**C**) SD.

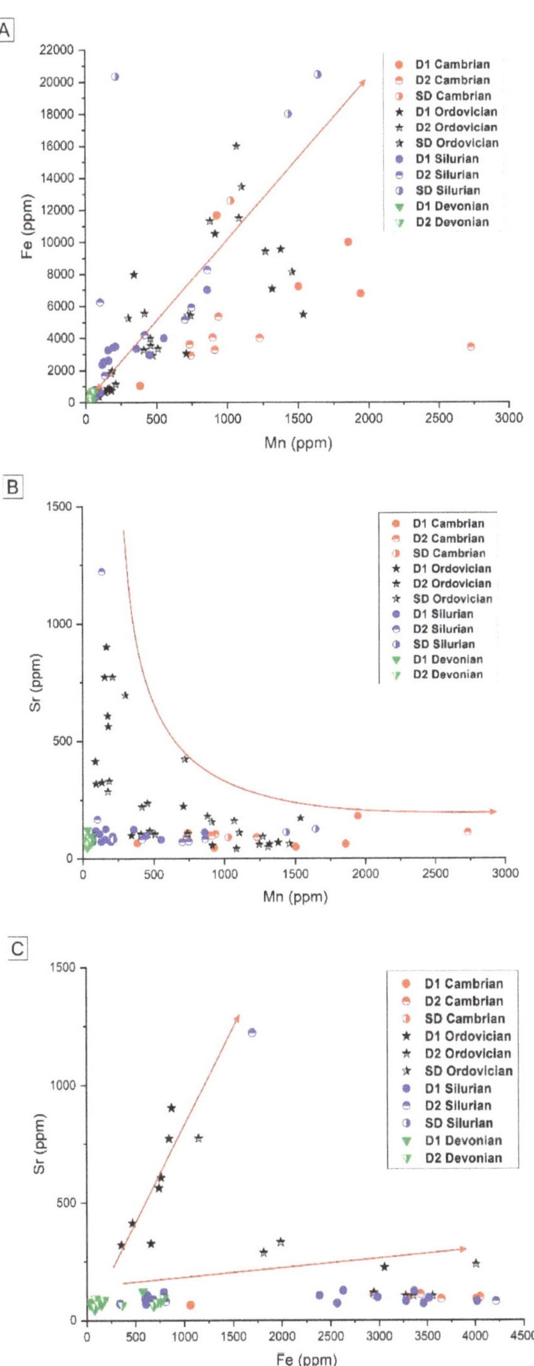

Figure 17. Cross plots of the investigated dolomites: (**A**) between Fe and Mn; (**B**) Sr vs. Mn concentrations, and (**C**) Sr vs. Fe.

Samples of dolomite and fracture- and vug-filling calcite cement from different age groups were analyzed for REE content (Table 4). The ∑REE of Cambrian dolomite types and calcite cement is significantly higher than those in Ordovician, Silurian, and Devonian strata (Table 4) compared with the average of warm water brachiopods shown by Azmy et al. [70]. ∑REE of Devonian samples record the lowest values among the rest.

Replacive dolomite D1 and D2 sampled from Cambrian formations show different trends (Figure 18) compared to the pattern of warm water brachiopods shown by Azmy et al. [70], with a dominant-negative La anomaly and positive Ce anomalies (Figure 19).

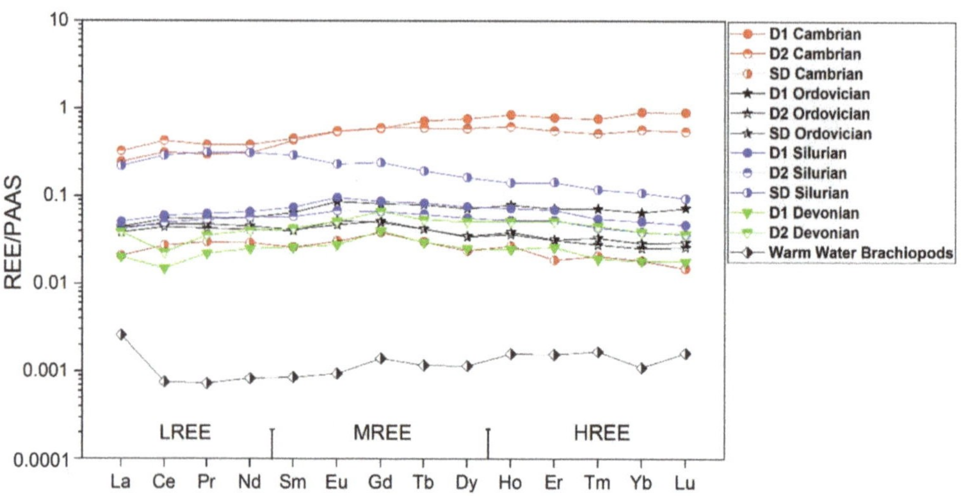

Figure 18. Post-Archean Australian Shale (PAAS) normalized rare-earth element (REE) pattern for average values of fine (D1) and medium (D2) crystalline dolomite, and saddle dolomite (SD) samples, compared with average PAAS normalized REE pattern of water brachiopods [70].

Table 4. Summary of the REE (ppm) composition of dolomite and calcite components in the studied successions.

Age-Phase	Stats.	La	Ce	Pr	Nd	Sm	Eu	Gd	Tb	Dy	Ho	Er	Tm	Yb	Lu	ΣREE
Cambrian-D1 (n = 5)	Avg.	10.96	27.92	3.04	11.58	3.01	0.67	3.62	0.65	4.11	0.90	2.45	0.35	2.77	0.40	72.43
	Stdev.	11.60	27.91	2.69	9.58	2.13	0.48	2.60	0.53	3.68	0.82	2.26	0.34	2.80	0.40	59.82
	Max.	33.20	80.29	7.84	27.95	5.25	1.26	6.71	1.48	10.10	2.26	6.16	0.91	7.51	1.08	170.43
	Min.	0.72	1.86	0.27	1.08	0.24	0.05	0.24	0.03	0.15	0.03	0.07	0.01	0.07	0.01	4.83
Cambrian-D2 (n = 7)	Avg.	14.68	38.36	3.95	14.53	3.16	0.68	3.67	0.54	3.19	0.66	1.73	0.24	1.74	0.24	87.38
	Stdev.	15.90	41.71	4.24	15.06	2.79	0.52	2.91	0.43	2.56	0.53	1.37	0.18	1.33	0.19	84.89
	Max.	52.30	138.00	13.90	49.38	8.87	1.53	8.42	1.45	8.95	1.88	4.81	0.64	4.66	0.66	282.20
	Min.	3.48	8.93	0.90	3.18	0.76	0.22	1.11	0.16	1.01	0.20	0.54	0.07	0.44	0.06	21.20
Cambrian-SD (n = 1)	Avg.															
	Stdev.															
	Max.	0.932	2.428	0.306	1.103	0.180	0.038	0.235	0.027	0.127	0.029	0.058	0.009	0.056	0.007	5.533
	Min.	0.932	2.428	0.306	1.103	0.180	0.038	0.235	0.027	0.127	0.029	0.058	0.009	0.056	0.007	5.533
Ordovician-D1 (n = 13)	Avg.	2.01	4.88	0.57	2.13	0.44	0.11	0.49	0.07	0.38	0.08	0.22	0.03	0.19	0.03	11.63
	Stdev.	3.04	8.62	1.06	4.15	0.96	0.23	1.05	0.16	0.93	0.20	0.54	0.08	0.48	0.08	21.56
	Max.	12.41	34.49	4.24	16.42	3.74	0.89	4.12	0.64	3.60	0.78	2.07	0.28	1.85	0.28	85.79
	Min.	0.20	0.45	0.06	0.21	0.03	0.01	0.05	0.01	0.03	0.01	0.01	0.00	0.01	0.00	1.07
Ordovician D2 (n = 16)	Avg.	1.68	3.85	0.43	1.52	0.27	0.06	0.30	0.04	0.18	0.04	0.09	0.01	0.07	0.01	8.54
	Stdev.	1.31	3.37	0.39	1.36	0.25	0.05	0.28	0.03	0.16	0.03	0.08	0.01	0.06	0.01	7.36
	Max.	5.62	14.02	1.60	5.65	1.05	0.23	1.17	0.15	0.70	0.14	0.34	0.05	0.24	0.04	30.99
	Min.	0.44	0.83	0.09	0.36	0.06	0.02	0.08	0.01	0.03	0.01	0.02	0.00	0.02	0.00	1.98
Ordovician SD (n = 6)	Avg.	1.97	4.26	0.49	1.69	0.29	0.06	0.32	0.04	0.19	0.04	0.10	0.02	0.09	0.01	9.55
	Stdev.	0.57	2.05	0.28	0.97	0.18	0.02	0.16	0.02	0.11	0.02	0.05	0.01	0.04	0.01	4.36
	Max.	3.08	7.75	0.98	3.28	0.61	0.10	0.61	0.09	0.40	0.08	0.21	0.03	0.15	0.03	17.41
	Min.	1.17	2.24	0.19	0.59	0.08	0.03	0.15	0.02	0.09	0.02	0.06	0.01	0.05	0.01	5.43

Table 4. *Cont.*

Age-Phase	Stats.	La	Ce	Pr	Nd	Sm	Eu	Gd	Tb	Dy	Ho	Er	Tm	Yb	Lu	ΣREE
Silurian D1 (n = 16)	Avg.	2.30	5.32	0.64	2.49	0.52	0.12	0.53	0.07	0.40	0.08	0.21	0.03	0.16	0.02	12.89
	Stdev.	1.07	2.84	0.35	1.38	0.31	0.07	0.30	0.04	0.24	0.04	0.12	0.01	0.09	0.01	6.84
	Max.	4.47	11.58	1.35	5.49	1.19	0.26	1.15	0.17	0.89	0.17	0.43	0.05	0.34	0.05	27.60
	Min.	0.78	1.42	0.15	0.52	0.10	0.02	0.10	0.01	0.06	0.01	0.04	0.00	0.03	0.00	3.25
Silurian D2 (n = 9)	Avg.	1.89	4.43	0.55	2.15	0.41	0.08	0.41	0.06	0.30	0.06	0.16	0.02	0.12	0.02	10.65
	Stdev.	1.32	3.34	0.43	1.67	0.31	0.06	0.31	0.04	0.23	0.04	0.12	0.02	0.08	0.01	7.95
	Max.	3.64	9.45	1.23	4.86	0.88	0.18	0.87	0.12	0.65	0.12	0.36	0.05	0.23	0.03	22.67
	Min.	0.28	0.50	0.05	0.23	0.05	0.01	0.05	0.01	0.04	0.01	0.02	0.00	0.02	0.00	1.30
Silurian SD (n = 4)	Avg.	8.84	23.37	2.95	10.78	1.87	0.27	1.36	0.16	0.83	0.14	0.41	0.05	0.30	0.04	51.37
	Stdev.	6.96	17.27	2.13	7.40	1.26	0.17	0.89	0.10	0.52	0.09	0.26	0.03	0.20	0.03	37.09
	Max.	19.39	47.98	5.92	20.35	3.42	0.44	2.30	0.27	1.34	0.23	0.69	0.08	0.52	0.06	102.93
	Min.	0.36	0.77	0.09	0.39	0.07	0.02	0.04	0.01	0.05	0.01	0.02	0.00	0.02	0.00	1.85
Devonian D1 (n = 13)	Avg.	0.91	1.34	0.23	0.95	0.18	0.03	0.24	0.03	0.14	0.03	0.08	0.01	0.06	0.01	4.17
	Stdev.	1.42	2.06	0.39	1.67	0.29	0.06	0.33	0.04	0.20	0.04	0.10	0.01	0.06	0.01	6.60
	Max.	4.80	7.60	1.35	5.77	0.98	0.18	0.97	0.11	0.64	0.12	0.33	0.04	0.23	0.03	22.88
	Min.	0.11	0.17	0.02	0.07	0.01	0.00	0.03	0.00	0.02	0.00	0.01	0.00	0.01	0.00	0.44
Devonian D2 (n = 14)	Avg.	2.01	2.23	0.40	1.65	0.32	0.07	0.43	0.05	0.30	0.06	0.18	0.02	0.13	0.02	7.80
	Stdev.	2.37	2.46	0.49	2.07	0.40	0.09	0.47	0.06	0.35	0.07	0.20	0.02	0.14	0.02	9.16
	Max.	7.20	7.62	1.48	6.15	1.18	0.25	1.32	0.18	1.03	0.20	0.57	0.06	0.42	0.05	27.72
	Min.	0.14	0.20	0.03	0.10	0.03	0.00	0.03	0.00	0.02	0.00	0.01	0.00	0.01	0.00	0.56

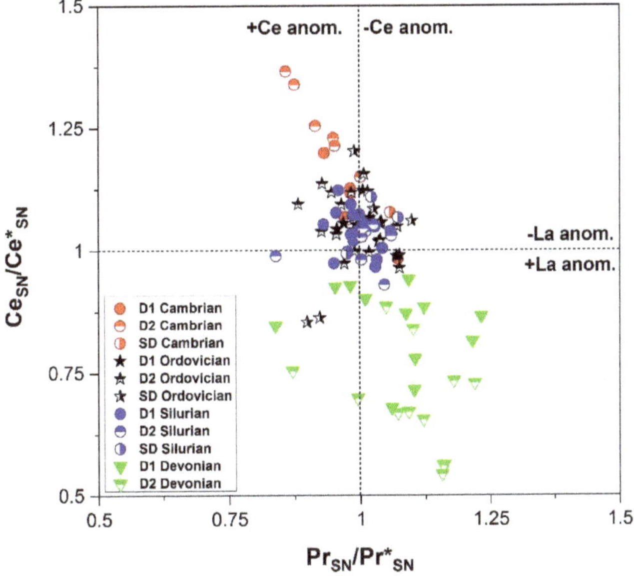

Figure 19. Cross plot Ce $(Ce/Ce^*)_{SN}$–La $(Pr/Pr^*)_{SN}$ anomaly of D1, D2, and SD samples of different age groups.

Both dolomite D1 and D2 (Figure 18) sampled from Ordovician formations show different trends compared to the pattern of warm water brachiopods shown by Azmy et al. [70], with a negative La anomaly and both cases of positive and negative Ce anomalies (Figure 19). Both D1 and D2 exhibit higher average ΣREE (11.63 ± 21.56 ppm, and 8.54 ± 7.36 ppm, respectively) compared with the average of warm water brachiopods shown by Azmy et al. [70]. Saddle dolomite cement (SD) patterns show lower average ΣREE (9.55 ± 4.36 ppm) than D1 but slightly higher than D2. It also shows, similarly to D1 and D2, slightly different patterns compared to warm water brachiopod's pattern with both negative La and Ce anomalies.

Replacive dolomite D1 and D2 sampled from Silurian formations show different trends compared to the pattern of warm water brachiopods shown by Azmy et al. [70] (Figure 18), with a minor negative La anomaly and both cases of positive and negative Ce anomalies (Figure 19). Brachiopods, D1, and D2 patterns with both negative La and Ce anomalies as well as a slight negative Eu anomaly (Figure 19).

Both fine and medium dolomite matrix samples (D1 and D2, respectively) from Devonian formations have shale-normalized patterns sub-parallel to those of modern warm water brachiopods from Azmy et al. [70]. They both show the typical characteristics of seawater patterns such as depletion of LREE over HREE and slightly positive La and Gd anomalies (Figure 19).

There is a subtle difference in Cambrian and Ordovician dolomite vs. Silurian and Devonian samples in terms of Sm/Yb vs. Eu/Sm (Figure 20A), La/Yb vs. La/Sm (Figure 20B), Y/Ho vs. Sm/Yb (Figure 20C) and Y/Ho vs. Eu/Sm (Figure 20D) ratios. For example, higher Sm/Yb ratios (Figure 20A), lower Sm/Yb, Eu/Sm (Figure 20C,D) in the Cambrian and Ordovician dolomites than Silurian and Cambrian dolomites.

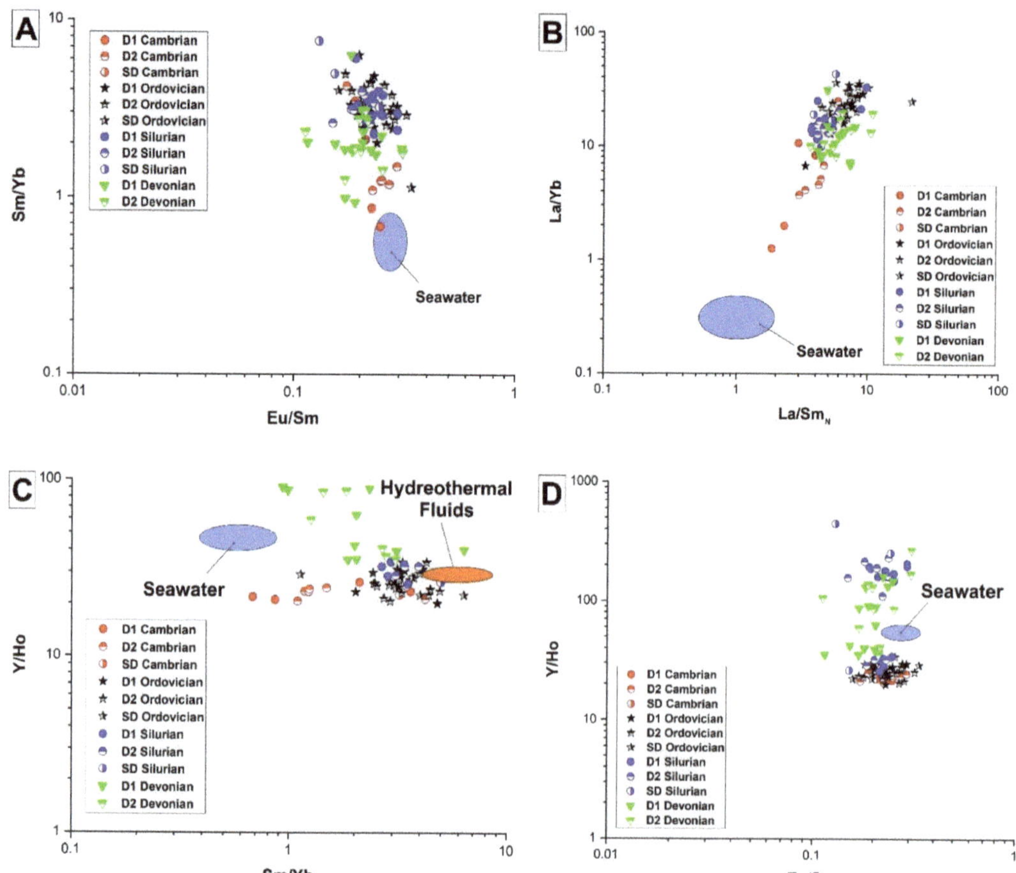

Figure 20. (**A**) Plots of Sm/Yb vs. Eu/Sm; (**B**) La/Yb vs. La/Sm_N; (**C**) Y/Ho vs. Sm/Yb, and (**D**) Y/Ho vs. Eu/Sm of dolomite samples from different age groups.

4.2.3. Fluid Inclusions Microthermometry

Microthermometric measurements of primary fluid inclusions from D2, SD, and BKC from each age group provide melting and homogenization temperatures (T_h) and estimates of salinity (cf. [61,71] for each phase (Table 5 and Figure 21). Primary fluid inclusions within D1 crystals were too small to measure (cf. [71]). However, monophase fluid inclusions are present in this dolomite. For calcite, measurements were taken for coarse crystalline cements (BKC) in vugs and fractures. The measured T_h provides an estimate of minimum entrapment temperatures. The low first melting temperatures and ice-melting temperatures suggest that the fluid within the fluid inclusions may approximate a H_2O-NaCl-$CaCl_2$ system. The examined inclusions are parallel to crystal facets and the vapor bubble size for the two-phase (liquid-vapor) fluid inclusions. There is a notable overlap in T_h and salinity values among all dolomite types in Cambrian, Ordovician and Silurian samples (Figure 21), showing high Th and salinity. However, lower T_h and salinity values are recorded for D2 in the Devonian samples. Fluid inclusions within D2 from Cambrian crystals yielded T_h that range from 75° to 132 °C (102.9 °C, n = 19), with salinity estimates based on final ice-melting temperatures (T_{mice}) that correspond to salinities range from 23.2 to 27.2 wt.% (25.4 wt.%, n = 10). Fluid inclusions in SD crystals have T_h ranging from 111° to 156 °C (131.6 °C, n = 5). Measurable 2-phase fluid inclusions in BKC giving T_h ranging between 85° and 141 °C (108 °C, n = 11), and salinities from 22.1 to 23.6 wt.% (22.9 wt.%, n = 2).

Table 5. Summary of the fluid inclusion data of the studied successions.

Age	Host Mineral	Size (μm)	T_h (°C)	Salinity
Cambrian	D2			
	n.	19	19	10
	Avg.	7.26	102.89	25.42
	Stdev.	2.88	17.05	1.32
	Max.	17.00	132.00	27.20
	Min.	4.00	75.00	23.20
Cambrian	SD			
	n.	5	5	0
	Avg.	8.40	131.60	
	Stdev.	1.96	20.05	
	Max.	10.00	156.00	
	Min.	6.00	111.00	
Cambrian	BKC			
	n.	11	11	2
	Avg.	11.00	108.00	22.85
	Stdev.	4.24	22.24	0.75
	Max.	20.00	141.00	23.60
	Min.	6.00	85.00	22.10
Ordovician	D2			
	n.	41	41	12
	Avg.	8.20	96.34	26.46
	Stdev.	2.37	17.08	3.39
	Max.	13.00	132.00	30.50
	Min.	4.00	66.00	21.60
Ordovician	SD			
	n.	10	10	8
	Avg.	9.00	115.20	23.34
	Stdev.	2.05	16.28	2.32
	Max.	12.00	152.00	25.90
	Min.	6.00	89.00	20.80

Table 5. *Cont.*

Age	Host Mineral	Size (μm)	T_h (°C)	Salinity
Ordovician	BKC			
	n.	27	27	7
	Avg.	11.77	109.67	27.19
	Stdev.	6.27	16.75	1.87
	Max.	22.00	153.00	30.30
	Min.	5.00	68.00	24.80
Silurian	D2			
	n.	42	42	15
	Avg.	3.52	93.45	23.10
	Stdev.	2.05	16.20	0.87
	Max.	10.00	134.10	25.21
	Min.	1.00	49.70	15.17
Silurian	SD			
	n.	53	53	29
	Avg.	6.02	123.51	28.54
	Stdev.	2.83	16.44	1.74
	Max.	15.00	193.40	32.63
	Min.	2.00	101.20	25.21
Silurian	BKC			
	n.	53	53	25
	Avg.	5.70	129.18	27.68
	Stdev.	3.31	31.92	2.55
	Max.	18.00	194.30	32.33
	Min.	2.00	70.20	21.40
Devonian	D2			
	n.	29	29	12
	Avg.	3.34	83.79	20.74
	Stdev.	1.67	8.26	0.90
	Max.	6.00	102.30	21.82
	Min.	1.00	69.90	18.96
Devonian	BKC			
	n.	48	48	25
	Avg.	5.73	202.04	20.75
	Stdev.	2.33	27.83	2.17
	Max.	12.00	255.10	23.95
	Min.	2.00	146.20	15.17

Fluid inclusions within D2 from Ordovician samples yielded T_h that range from 66° to 132 °C (96.3 °C, n = 41), with salinity, estimates that range from 21.6 to 30.5 wt.% (26.5 wt.%, n = 12; Table 5). Fluid inclusions in BKC giving T_h ranging between 68° and 153 °C (109.7 °C, n = 27), and salinities from 24.8 to 30.3 wt.% (27.2 wt.%, n = 7).

In Silurian sample T_h values for D2 are ranging between 49.7 ° and 134.1 °C (93.4 °C, n = 42), and salinities from 15.1 and 25.2 (23.1, n = 15). Fluid inclusions in SD crystals have T_h ranging from 101.2° to 193.4 °C (123.5 °C, n = 53), and salinities from 25.2 to 32.6 (28.5 wt%, n = 29). Measurable 2-phase fluid inclusions in BKC giving T_h ranging between 70.2° and 194.3 °C (129.1 °C, n = 53), and salinities from 21.4 to 32.3 wt.% (27.6 wt.%, n = 25).

Fluid inclusions within D2 from Devonian samples yielded T_h that range from 69.9° to 102.3°C (83.7 °C, n = 29), with salinity range from 18.9 to 21.8 wt.% (20.7 wt.%, n = 12). For BKC higher T_h values than samples from other age groups are measured and vary between 146.2° and 255.1 (202.0 °C, n = 48), and lower salinity values vary between 15.1 and 23.9 wt.% (20.7 wt.%, n = 25).

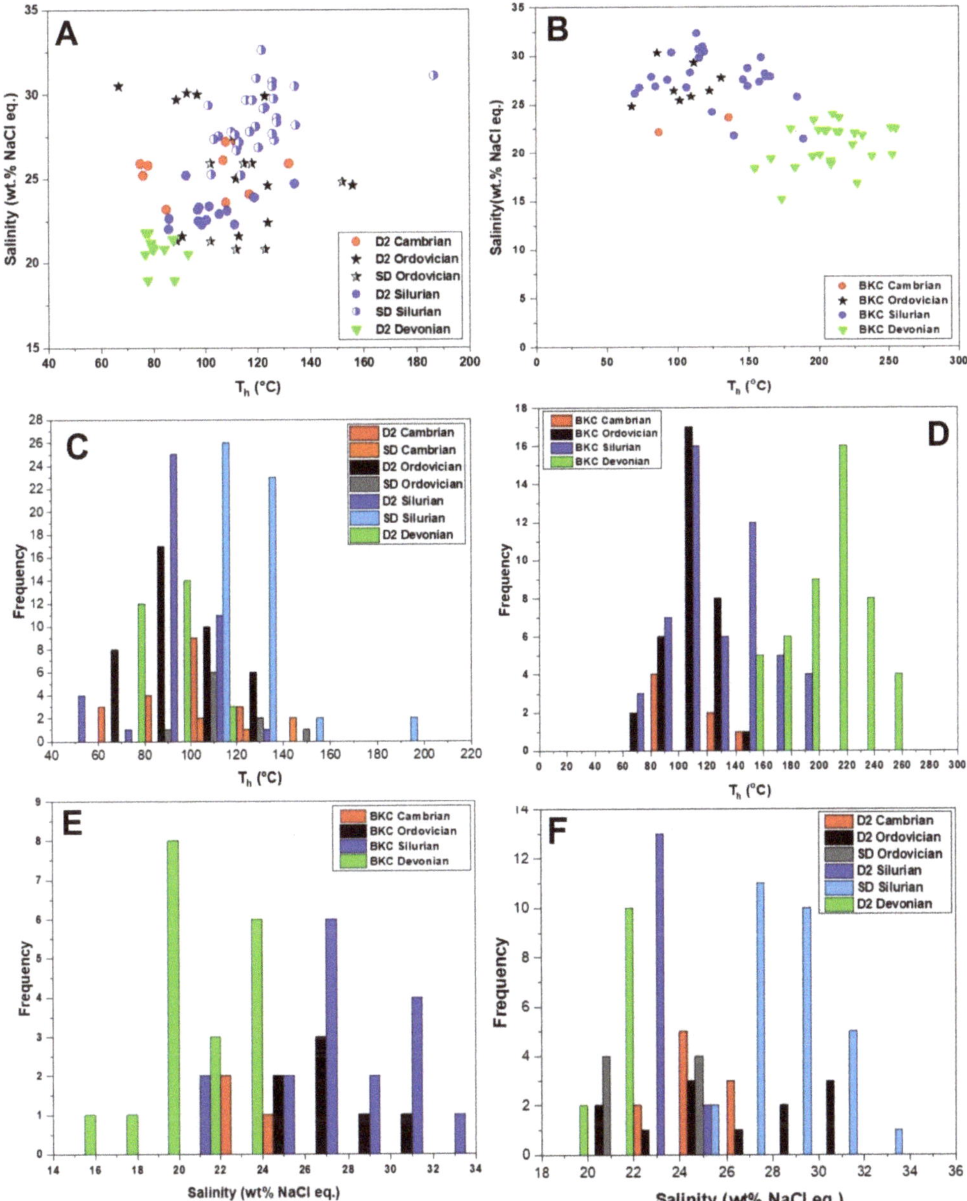

Figure 21. Cross plots of (**A**) Homogenization temperature vs. salinity of dolomite types; (**B**) calcite phases; (**C**) Histogram plot of T_h for dolomite samples; (**D**) Histogram plot of T_h for calcite samples; (**E**) Histogram plot of salinity of late calcite; and (**F**) Histogram plot of salinity for dolomite types.

5. Data Analysis and Discussion

Petrographic investigations, stable-isotope, elemental analyses, and fluid-inclusion microthermometry provide important clues regarding dolomitization and other diagenetic processes, fluid chemistry, and the history of fluids migration in the sedimentary basin. The Paleozoic successions in the Huron Domain of Michigan Basin were affected by early (marine to shallow eogenetic zone; [72]) and deeper (mesogenetic zone) diagenetic processes [72]. These include early calcite cementation, silicification, partial dissolution, early dolomitization and later dolomitization, recrystallization, fracturing, stylolitization, late calcite, and sulphate cementation. There are similarities in the diagenetic processes in Cambrian and Ordovician carbonates and Silurian and Devonian carbonates which allow us to consider and discuss these carbonates together (Figure 4).

5.1. Constraints from Petrography-Paragenetic Sequence

The paragenetic sequence of the main diagenetic events in the studied Paleozoic successions (Figure 4), which was constructed originally based on textural relationship and geochemical and fluid inclusion evidence, provides a useful platform for further discussion of the data obtained in this study.

As reviewed in Section 2 above, the depositional environments of the Cambrian and Ordovician carbonates encompass various lithofacies assemblages, including supratidal, lagoonal to subtidal environments (e.g., [47,55]). The diverse lithofacies changes observed in these formations are directly controlled by relative sea-level changes and sources of detritus. These were evident especially in the Cambrian strata where mixed siliciclastic-carbonate facies are common. Ultimately, these depositional processes also affected the diagenetic changes encountered in these sediments during shallow and deeper burial stages. In carbonate-rich units, the bioclastic grainstone facies has been influenced by early stabilization of calcitic/aragonitic components and dolomite replacement at shallow burial (D1; Figure 5B) to be succeeded by medium to coarse crystalline dolomite (D2; Figure 5B–D,I), possibly at an intermediate burial. Minor presence of saddle dolomite (SD) occluding vugs postdates D2 but predates BKC, which is occluding fractures; both of which probably formed later than D2 in the intermediate burial stage (Figure 4; Figure 5E). In the ooid grainstone lithofacies, which is extensively silicified (Figure 5G), the early isopachous calcite cement was silicified. However, the silicified ooids are well preserved and show evidence of very early diagenetic, the pre-compaction origin of these allochems. The siliciclastic-dominated facies were cemented earlier by calcite cement later to be compacted with the creation of some minor fractures occluded by equant calcite cement, which was partly dolomitized. The basal Cambrian sandstone is assumed to have been a major migratory path for diagenetic fluids [73,74]. Strata in the eastern region of the basin pinches out towards the margins. These thin fingers of clastic material which have large amounts of bioclastic material increase the likelihood for extensive dolomitization of a thin layer.

In the Ordovician formations (e.g., Coboconk Formation) diagenesis was initiated by early cementation that produced the nodular facies on the seafloor. Burial and compaction affected the formation of these nodules and may have created hairline fractures, later occluded by dolomite. Dolomitization (D1; Figure 6) occurred by selective early replacement of skeletal components such as bryozoans and echinoderms and the micritic matrix, as well as, pervasively replacing the intra-nodule matrix during shallow burial. These dolomites were crosscut by hairline fractures and stylolites during a burial at a later stage, suggesting the formation of D1 at a shallow burial. Silicification, which occurs in some facies of the Coboconk Formation occurred prior to dolomitization as evidenced by silica replacing the matrix and fossils that later were selectively dolomitized. In the Gull River Formation, a similar paragenetic sequence to the Coboconk Formation is observed (Figure 4). Initially, dolomitization proceeded with fine crystalline (D1) and medium to coarse crystalline matrix replacement dolomite (D2). The minor saddle dolomite (SD) cement postdates earlier dolomite formation and occludes vugs and fractures and is also postdated by late, blocky calcite cement (BKC), which is occluding vugs and fractures (Figure 6). Other compaction features such as stylolites postdate or sometimes appear to be contemporaneous with the medium to coarse crystalline dolomite (D2; Figure 6C). In both the Coboconk and Gull River formations, the intra and internodule dolomites typically present as petrographically distinct. These distinctions reflect differences in the underlying substrate that was originally dolomitized and relate to permeability, not fluid chemistry, or timing (notwithstanding some evolution of brine chemistry over time). An early, shallow burial origin is suggested for the fine crystalline (D1) dolomite as evidenced by the selective replacement of this dolomite, its shape and size, and it predates stylolite formation, consistent with a shallow burial diagenetic environment. In contrast, the medium and coarse crystalline (D2) dolomite may have formed at medium burial stages as evidenced by its increasing crystal size, subhedral and anhedral shape (e.g., [26]), and close association sometimes with compaction features, such as stylolites. The stylolites observed in the samples all occur at ~90° to the fractures. This relationship may be explained by compressive stress, imposed by the Taconic Orogeny, which was occurring concurrently with deposition. These stylolites may have acted as conduits for Mg-rich fluids (e.g., [75]).

In contrast, the depositional environments of the Silurian and Devonian successions encompass various lithofacies assemblages, including shallow marine ramp and deeper water subtidal lithofacies dominated by limestone and dolostone lithofacies with intercalations of shales and evaporite deposits in both age groups (e.g., [47,76,77]). In the Silurian and Devonian carbonates early calcite cementation (e.g., ISC and SXC; Table 1) and micritization initiated on the seafloor to be followed during burial by occlusion or porosity by later calcite cementing vugs and fractures (e.g., DC, and BKC; [78]). BKC postdates saddle dolomite (SD; Figure 4). Evaporite minerals, such as gypsum and anhydrite represent the final stage of cementation in the Silurian rocks. Dolomitization commenced early (D1) formed replacing calcitic and aragonitic allochems and muddy calcitic matrix [3]. This dolomite may have formed at a shallow burial, in a reflux evaporative setting (e.g., [33,79,80]). This is supported by its fine crystalline crystal size [68] and the cross-cutting relationship with compactional features, such as dissolution seams and stylolites which crosscut this dolomite. This dolomitization phase is followed by the formation of medium to coarse crystalline replacive dolomite (D2) at the medium burial stage. The increase in crystal size of D2 may reflect dolomitization of coarser lithofacies (such as grainstones) and/or recrystallization of precursor dolomite matrix (D1; [23]). In the Silurian carbonates, only saddle dolomite (SD) is observed occluding fractures, postdating D1 and D2 but predating blocky calcite (BKC; Figure 7). This dolomite formed possibly in an intermediate burial setting.

Evaporite minerals (gypsum and anhydrite; Figure 10) represent the last diagenetic phases and form in a deeper burial realm. These minerals occlude fractures and vugs in Cambrian, Ordovician, and Silurian, carbonates.

5.2. *Controls of Dolomitization-Geochemical Evidence*

5.2.1. Stable and Radiogenic Isotopes

Cambrian:

The stable isotopic composition of D1, D2, and SD in Cambrian rocks (Figure 11) are notably lighter than the postulated dolomite precipitated in equilibrium with marine waters (e.g., [69]). These values are also distinct from other age groups (Figure 11). There is little isotopic fractionation of $^{13}C/^{12}C$ with temperature, and hence the $\delta^{13}C$ values of these dolomite reflect isotopic fractionation of C related to fluids that are enriched in lighter CO_2 originated from oxidized carbon, which can be related to seawater-sourced, reduced, bacterial sulfate reduction process for D1 and hydrocarbon sources for D2 and SD [2].

However, the $\delta^{18}O$ values of the dolomites reflect both the temperature and the composition of the parent fluids, and thus the departure from equilibrium values reflects precipitation of these dolomite under a different diagenetic regime [81]. The negative shift in $\delta^{18}O$ of D1 and D2 (Figure 11) signifies the effect of temperature increase during burial, change in fluid composition, or both [81]. Recrystallization of D1 with increasing temperature/change in fluid chemistry can also result in a negative shift in D1 (e.g., [23]). This recrystallization is also observed petrographically, as shown by the increasing crystal size of D1 and the presence of overgrowth zonation in D2 (Figure 5). The co-variant trend between carbon and oxygen isotopes in these dolomites reflect varied water-rock interaction with increasing burial (cf. [82]). Water rock interaction is also evidenced by the presence of more radiogenic $^{87}Sr/^{86}Sr$ ratios in D1 and D2 (Figures 13 and 15) in comparison to the postulated range for Cambrian seawater. This can be related to the presence of more siliciclastics, rich in clays and feldspars, in the Cambrian rocks as well as can be sourced from the Precambrian shield [74,83]. The same trend and comparable negative isotopic values are evident in late calcite cement (BKC; Figure 12) and thus reflecting similar diagenetic conditions during later fluid flow events for these carbonates unique to Cambrian paleohydrologic flow system. Similarly, BKC also shows more radiogenic Sr isotope ratios than the Cambrian seawater values (Figure 14).

Ordovician:

Most of the oxygen and carbon isotopic values of the Ordovician replacive dolomites (D1 and D2) show overlapping signatures with a departure by at least 3‰ from their coeval postulated seawater equilibrium values with respect to their age (Figure 11), especially for oxygen isotopes. This can be related to the recrystallization of D1 to D2 at elevated temperatures during the intermediate burial stage (Figure 4). C isotope values are closer to their seawater equilibrium values reflecting the buffering of this isotope with original carbonate values (e.g., [26,29]). However, $\delta^{13}C$ values of saddle dolomite show slight negative shifts compared to the postulated equilibrium values. This may reflect an oxidized organic C source during burial. The $^{87}Sr/^{86}Sr$ ratios of many samples of D1 and D2 show comparable and/or slight radiogenic signature to seawater values (Figure 13; [69]) with the exception of few dolomites that show highly radiogenic rations (Figures 13 and 14). These dolomite samples occur along dissolution seams and thus reflecting a radiogenic fluid source of Sr, perhaps from Cambrian siliciclastic fluid. In contrast, the isotopic composition of early calcite cement shows preservation of the original seawater signature (Figure 12), while late calcite cement shows some negative trend in both oxygen and carbon isotopes relative to the postulated values in Ordovician calcite. This reflects the diagenetic progression and precipitation of this cement from fluids modified at the burial. This is also shown by $^{87}Sr/^{86}Sr$ ratios of BKC which show seawater signature (Figure 14).

Silurian and Devonian:

The isotopic data of the overlying Silurian and Devonian carbonates show overlaps between $\delta^{13}C$ and $\delta^{18}O$ values (Figure 11). Dolomite samples from both age groups show similar $\delta^{13}C$ values compared to those of carbonates deposited in equilibrium with seawater of respective age except for few D1 samples from the Silurian carbonates that show a slight negative shift, which can be related to a depleted carbon source related to bacterial sulfate reduction at the shallow burial [84]. This is supported by the presence of minute framboidal pyrite crystals embedded within RD1 (this study and Tortola et al. [3]). Also, few BKC samples from the Silurian carbonates show a negative shift from the equilibrium values (Figure 12). This can be related to an oxidized organic carbon source, possibly related to interaction with hydrocarbons (e.g., [2]). However, $\delta^{18}O$ values of D1 show evidence of dolomite recrystallization as these values depart from their respective age equilibrium values. This negative shift in oxygen isotopes also reflects increasing temperature and/or changes in the diagenetic fluid composition with increasing burial (cf., [3,26]). $^{87}Sr/^{86}Sr$ ratios of many samples, both dolomite and calcite, from these age groups cluster close to their coeval seawater composition (Figure 13). However, a few dolomite samples show slight enrichment of radiogenic Sr ratios. This can be related to the presence of modified seawater fluids upon burial.

5.2.2. Major, Minor, and REE Elements

The average values of $CaCO_3$ in all analyzed replacement and cement dolomite types show that all dolomite types are non-stoichiometric, consistently calcium-rich (Table 4, average $CaCO_3$ range from 54.3 to 61.8 mol% for D1 and D2, and from 57.7 to 63.3 mol% for SD, respectively; Figure 16). This suggests that the precursor mineralogy of these dolomites is of high-Mg calcites and aragonite and the stabilization of these minerals to dolomite occur in semi-closed diagenetic systems with less fluid mobility and limited supply of Mg ions (cf., [18,85]). Hence, dolomitizing fluids reflect a lowering Mg/Ca ratio in these fluids which encourages Ca-rich, non-stoichiometric dolomite to form. In general,

Mn and Fe contents in the Cambrian and Ordovician dolomites, which are indicators of redox conditions during dolomitization, are much higher than the Silurian and Devonian dolomites (Table 3) reflecting a decrease in oxidizing conditions in the older successions. There is also a covariant trend between Mn and Fe for these dolomites (Figure 17) reflecting the geochemical affinity of these two elements [86]. However, the average of Sr content in all dolomites does not show large variations among the investigated dolomite types (vary between 83 and 253 ppm, Table 3) and are comparable to those of other ancient dolomites but lower than the estimated values for dolomite in equilibrium with seawater (~500–800 ppm, (e.g., [87]). However, many samples from the Ordovician dolomites show high Sr content and covariant trends with Fe and Mn (Figure 17), perhaps reflecting the aragonitic precursor mineralogy of these dolomites. Such a relationship is not apparent in dolomites from other age groups.

Rare earth elements (REEs) have been commonly utilized for the reconstruction of a paleoenvironmental history of ancient seawater [88–90], diagenetic alteration, and the temperature of fluids precipitating carbonate minerals [70]. REE can also provide information about the relative water depth and the proximity to hydrothermal sources [91]. Some workers indicate that carbonates might preserve their REE compositions and/or patterns during diagenetic processes [89,92]. However, many studies have revealed that consequent diagenetic alteration modifies REE concentrations and/or patterns [70,93]. Therefore, the redistribution of REE in carbonates is an indicator that can be utilized to ascertain the potential contributing factor to observed REE concentrations in diagenetic minerals, especially in high water/rock ratio diagenetic systems [70]. The rare earth elements generally exist in the stable trivalent (3+) oxidation state, whereas Eu (Eu^{2+}, Eu^{3+}) and Ce (Ce^{3+}, Ce^{4+}) exhibit distinct behavior due to the redox state of the natural system (e.g., [93]). Ce (Ce^{3+}) can be oxidized to Ce (Ce^{4+}) in marine systems under surface conditions, whereas high temperature and/or high reducing conditions cause Eu (III) to be reduced to Eu (II) [65,94]. These characteristics provide the utilization of REEs as tracers of chemical processes during the diagenetic history of carbonates [89,95–97].

The $\sum$REE of Cambrian dolomite types and calcite cement (Table 3) is significantly higher than those in Ordovician, Silurian, and Devonian strata when compared with the average of warm water brachiopods shown by Azmy et al. [70]. $\sum$REE of Devonian samples record the lowest values among those examined in this study. The higher concentration of $\sum$REE of Cambrian dolomites signify that the dolomitizing fluids contained higher concentrations of REE, likely related to the interaction of these fluids with basement rocks.

Replacive dolomite D1 and D2 sampled from Cambrian formations show different trends (Figure 18) compared to the pattern of warm water brachiopods shown by Azmy et al. [70], with a dominant-negative La anomaly and positive Ce anomalies (Figure 19). These dolomites may have formed under suboxic to anoxic conditions [70]) from early crustal fluids highly enriched in REE particularly MREE and HREE relative to dolomites from other age groups. Saddle dolomite from the Cambrian rocks show evidence of different fluids that produced D1 and D2 are likely related to hydrothermal fluids that originated from deeper sources and transported via fractures and faults [2]. The flat Ce* anomaly, slightly positive Eu*, flat LREE over HREE, and sight MREE enrichment with low La/Sm in saddle dolomite reflect higher temperature fluids of seawater origin under low oxic conditions modified by siliciclastic contamination.

Both D1 and D2 (Figure 18) sampled from Ordovician formations show different trends compared to the pattern of warm water brachiopods shown by Azmy et al. [70], with a flat to slight Eu anomaly, negative La anomaly, and both cases of positive and negative Ce anomalies (Figure 19). Both, D1 and D2 exhibit higher average ΣREE (11.59 ± 22.44 ppm, and $8.54 + 7.36$ ppm, respectively) compared with the average of warm water brachiopods shown by Azmy et al. [70]. Saddle dolomite cement (SD) patterns show lower average ΣREE (9.55 ± 4.36 ppm) than D1 but slightly higher than D2. It also shows, similarly to D1 and D2, slightly different patterns compared to Warm water Brachiopod's pattern with both negative La and Ce anomalies. High Eu*, high La/YbN* and Y/Ho can indicate interaction with hydrothermal fluids [90].

Silurian replacive dolomites (D1 and D2) show different trends compared to the pattern of warm water brachiopods shown by Azmy et al. ([70]; Figure 18), with a minor negative La anomaly and both cases of positive and negative Ce anomalies (Figure 19). Brachiopods, D1 and D2 show patterns with both negative La and Ce anomalies as well as a slight negative Eu anomaly (Figure 18). These data suggest a seawater-dominated source of diagenetic fluids evolved during burial and increasing water/rock interaction [3]. The higher ΣREE and a more pronounced enrichment in LREE in SD and BK reflect a different source of REEs and/or involvement of saline fluids.

Both, fine and medium dolomite matrix samples (D1 and D2, respectively) from Devonian formations have shale-normalized patterns sub-parallel to those of modern warm water brachiopods from Azmy et al. [70]. They both show the typical characteristics of seawater patterns such as slight depletion of LREE over HREE and slightly positive La and Gd anomalies (Figure 19) potentially modified from the interaction of meteorically derived fluids and fluids of marine parentage. This can occur at near-surface, slightly reducing conditions. Although there is no evidence of karstification in the samples collected and analyzed in this study, a previous study [98] suggested that dissolution processes were mostly active in the shallow subsurface, usually <200 m depth in south-western Ontario. Thus, the possibility of meteoric water influence cannot be discounted completely.

There is a subtle difference in Cambrian and Ordovician dolomite vs. Silurian and Devonian samples in terms of Sm/Yb vs. Eu/Sm (Figure 20A), La/Yb vs. La/Sm (Figure 20B), Y/Ho vs. Sm/Yb (Figure 20C) and Y/Ho vs. Eu/Sm (Figure 20D) ratios. These differences may reflect changes in fluid composition, temperature, and/or redox conditions [99,100]. For example, Y/Ho vs. Sm/Yb ratios showing two possible fluids (marine and hydrothermal), and their interactions more evident in Cambrian and Ordovician dolomites. In contrast, the effect of seawater on the formation of Silurian and Devonian dolomites is more important. Other ratios document the presence of modified seawater as the primary fluid type in the formation of these dolomites [100].

5.2.3. Fluid Inclusions

Previous studies on fluid inclusions in Cambrian to Devonian dolomite and calcite in both central and marginal parts of the Michigan basin demonstrated fairly a wide range of T_h and salinity values (e.g., [2,3,32,33,101–103]). These investigations reported high T_h and salinity values in the central part of the basin compared to the margin of the basin, which some of them attributed to hydrothermal fluids which originated from the central part of the basin. Haeri-Ardakani et al. [33] further suggest a thermal anomaly related to the mid-continent rift under the Michigan Basin during late Devonian-Mississippian is the major source of heat for hydrothermal fluids.

The presence of single-phase fluid inclusions in D1 suggests very low temperatures, up to 50 °C [71]. However, the salinity-homogenization temperature plots (Figure 21A,B) for both dolomite (D2 and SD) and calcite (BKC) demonstrated a clear and subtle difference between age groups. For instance, higher T_h and salinity values are recorded in SD and some D2 in Ordovician and Silurian samples compared to Devonian dolomite. In Cambrian, D2 overlaps with Ordovician samples. In contrast, BKC in Devonian samples show the highest T_h and lowest salinity values compared to the rest (Figure 21B). High salinity and homogenization temperatures values are shown by Silurian dolomite and calcite. This reflects the effect of fluids sourced from saline brines related to evaporite dissolution in the Silurian rocks (e.g., [3,32]). The differences in homogenization temperature and salinity values in BKC (Figure 21) between Silurian and Devonian may reflect divergent fluid systems, such as a fluid that is characterized by high temperature and high salinity in the Silurian rocks and much higher temperature and lower salinity fluids in the Devonian rocks. These phenomena can also reflect the effect of Alleghenian Orogeny in Carboniferous time (Figure 1) as a factor that affected the fluid composition and sources.

In all analyzed D2 and SD dolomite and late BKC in the investigated age groups the high T_h values are significantly higher than the maximum burial temperatures suggested by other studies (e.g., [2,32,33,101,102]). This may suggest an involvement of hydrothermal fluids during the formation/recrystallization of dolomite and precipitation of late calcite cement. Increasing temperature of formation/recrystallization, as well as water/rock interaction, can explain the negative departures from equilibrium $\delta^{18}O$ values of dolomite.

5.3. Dolomite Recrystallization

Dolomite crystal shape, its boundaries, and its characteristics are directly related to the crystallization environment in which they formed [104]. Recrystallization of earlier formed dolomite (e.g., [23,105–107]) can involve change in texture, such as coarsening of crystal size or increase in nonplanar boundaries, change in structure (progressive ordering and strain), change in composition, which include isotopes, trace elements, stoichiometry, fluid inclusions, and zonation. It can also include a change in paleomagnetic properties [11].

Recrystallization of D1and D2 dolomites in the studied rocks has been invoked because of petrographic and geochemical data. Alteration of early formed dolomite is a common diagenetic process during progressive burial and/or increasing water/rock interactions (e.g., [8,108,109]). SEM and petrographic evidence for recrystallization of D1 include: (1) increase of crystal size (Figure 22); (2) lack of zonation (e.g., overgrowth) under CL, especially in D2; (3) lack of stoichiometry in D1 and D2 can be related to recrystallization of D1 to D2 by Ca-rich, warmer basinal fluids, perhaps under a semi-isolated diagenetic system (e.g., [3,13,110]). Geochemical and isotopic evidence for recrystallization with increasing burial include: (1) a negative shift of $\delta^{18}O$ from postulated values of marine dolomite of respective ages (Figure 11), (2) a slight enrichment of Sr isotopic ratios (Figure 13), (3) high T_h values in D2 (Figure 21), and (4) trace elements fractionation in D1 and D2 (Figure 17).

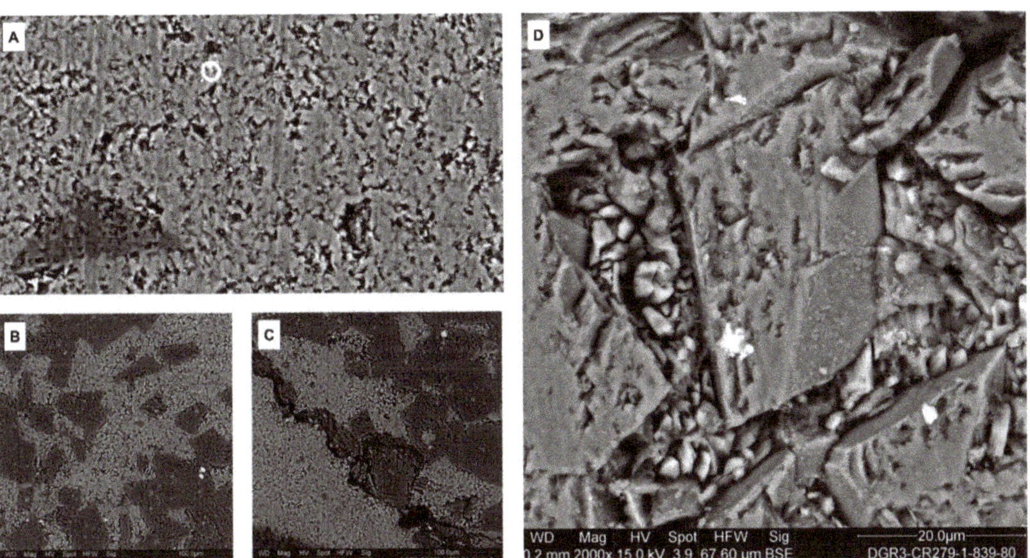

Figure 22. SEM and BSE images of D1 (**A**), D2 (**B**,**C**) and SD (**D**).

5.4. Evolution and Origin of the Diagenetic Fluids

Fluid flow in sedimentary basins can be linked to the tectonic evolution of the basin which can involve tectonic thrusting, sedimentary loading, uplift, and compression [4]. These fluids can transfer dolomitizing fluids as well as metals and hydrocarbons. Recent studies on diagenetic fluid flow in southern Ontario provided models on the fluid composition and its source(s). Coniglio and Williams-Jones [111], Middleton et al. [55] and Coniglio et al. [112] advocated that lateral and vertical movement of Mg-bearing fluids in Michigan Basin resulting from the transformation of clay minerals, originating from younger, older, or correlative argillaceous sediments from deeper in the Michigan and Appalachian basins towards the basin margin was considered as a mechanism for migration of dolomitizing fluids. Yoo et al. [113] also considered compaction driven flow to explain the mass and heat from the Michigan Basin. Haeri-Ardakani et al. [32,33] suggested that hot brines in the central part of the basin migrated through the basal Cambrian sandstone and ascended through a fracture network precipitating dolomite and late-stage calcite while carrying hydrocarbons.

Hobbs et al. [114] suggested two geochemical systems present on a regional scale; a shallow system (<200 m) below ground surface containing fresh through brackish waters and an intermediate to deep (>200 m) system containing brines associated with hydrocarbon reservoirs. Petts et al. [115] using microthermometric and isotopic data of secondary minerals in Cambrian and Ordovician strata suggested the presence of hydrothermal brines mixed with connate water. Al-Aasm and Crowe [2] in their investigations of dolomitization in the Cambrian and Ordovician carbonates in the Huron Domain identified two somewhat isolated digenetic fluid systems; an earlier Cambrian system and a later Ordovician system, both displaying unique geochemical attributes. Bouchard et al. [116] suggested that the fluids in the Ordovician low permeability sequences in the eastern part of Michigan Basin have originated as Ordovician seawater that was chemically overprinted by evaporated seawater that infiltered during the late Silurian and mixed with pre-existing brines from a deeper part of the hydrostratigraphic section. More recently, Tortola et al. [3] suggested the presence of a diagenetic fluid system that affected Silurian carbonates and was altered by salt dissolution post-Silurian and a Devonian fluid system that affected by the Alleghanian orogeny and resulted in a different fluid source and composition.

Petrographic relationships and isotope data can be used to infer changes in the composition of pore fluids in the Paleozoic successions during progressive diagenesis. Using the mean Th values and the range of $\delta^{18}O$ for D2 and SD from the investigated time frames, the $\delta^{18}O$ composition of the precipitating fluids was determined. The relationships among mineral $\delta^{18}O$ values, pore water $\delta^{18}O$ values, and temperature for diagenetic phases from the investigated carbonates are shown in Figure 23A,B. The suggested pathways for pore water evolution reflect changes in temperature, changes in fluid chemistry, or both. For dolomite types/generations (Figure 23A), there is a subtle difference in temperature and hence fluid composition between D2 and SD in Cambrian strata. The dolomitizing fluids for D2 reflect marine parentage, while for SD it is more enriched in $\delta^{18}O$ values suggesting the modified brines formed under higher temperatures. In contrast, there is a significant overlap in the $\delta^{18}O$ composition of dolomitizing fluids for Ordovician D2, SD, and Devonian D2 and a comparable range of formation/recrystallization temperatures. However, there is a measurable difference between D2 and SD in Silurian samples, as D2 formed under lower temperatures and comparable seawater $\delta^{18}O$ values but higher temperatures and more enriched values for SD formation. This can be related to an increase in salinity with burial, which can be linked to the Salina salt (dissolution) in the Upper Silurian, and the incursion of hydrothermal fluids during the formation of SD (e.g., [117]).

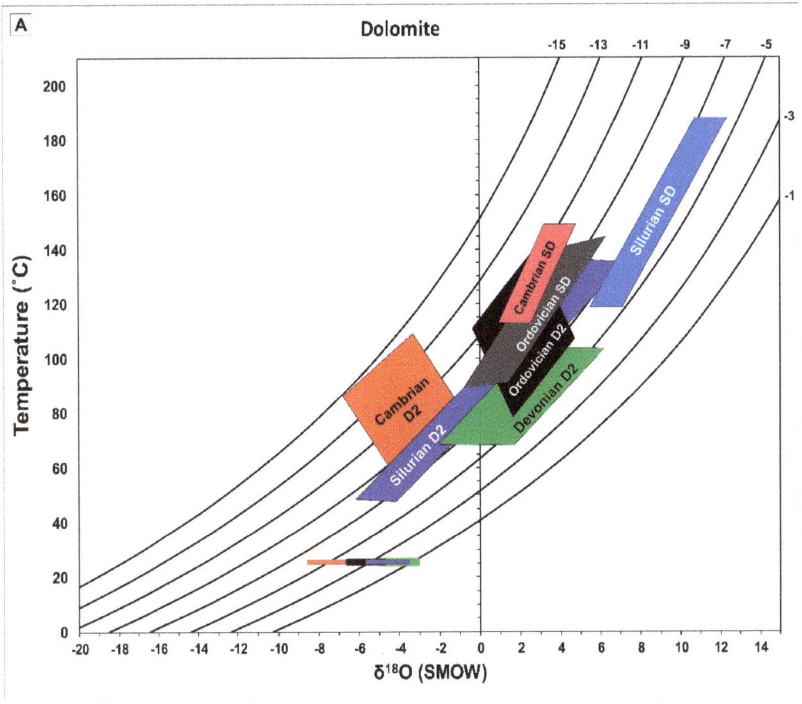

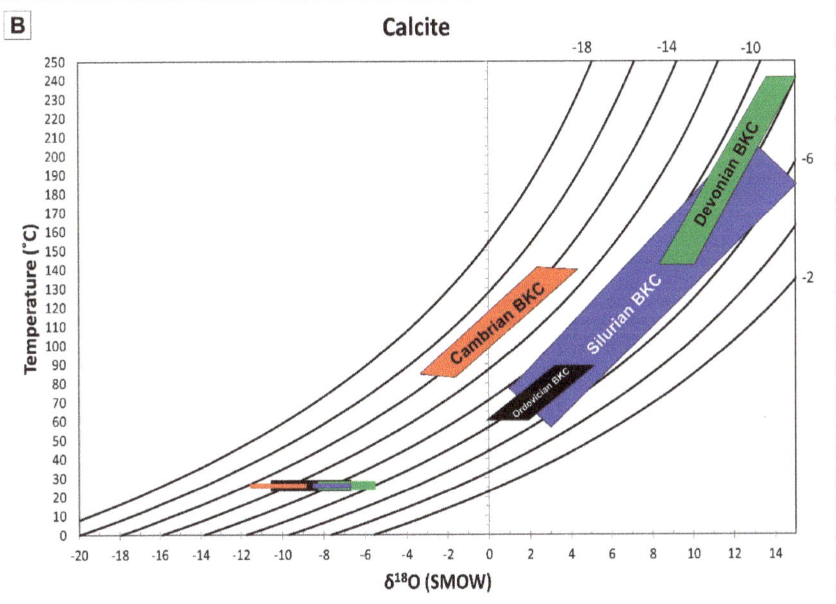

Figure 23. (**A**) Oxygen isotope values of dolomite the investigated formations plotted on the temperature-dependent, dolomite-water oxygen fractionation curve [118]. (**B**) Oxygen isotope values of calcite cement from the investigated formations plotted on the temperature-dependent, calcite-water oxygen fractionation curve [119].

The formation of blocky calcite (BKC) later in the diagenetic sequence reflects variations in formation temperatures and fluid composition in the investigated age groups (Figure 23B). All calcites formed at higher temperatures and deviate from the seawater composition of corresponding ages. However, similarities in $\delta^{18}O$ fluid composition between BKC and D2 in the Cambrian samples suggest the fluids could have originated from the Precambrian rocks below and migrated to the overlying porous Cambrian strata (cf. [74]). In contrast, the composition and temperatures of fluids that formed the late calcite in other age groups changed, becoming more enriched and thus reflecting evolved basinal brines [2]. The pathways of these fluids are shown by increasing both $\delta^{18}O$ values and temperature with time (from older to younger systems). Variations in pore fluid isotopic composition in the Paleozoic age groups suggest compartmentalization of these diagenetic fluids resulting in the formation and/or recrystallization of dolomite and calcite cementation. This phenomenon is also supported by radiogenic Sr isotopic ratios of these diagenetic phases as well as REE distribution of these carbonates (Figures 14 and 23).

Fluid compartmentalization in carbonate strata in the Huron Domain is also illustrated in Figure 24, which shows some of the important geochemical parameters used in this study vs. their stratigraphic position. There is a subtle difference in stable isotopic composition between different age groups, especially in the Cambrian samples. In all dolomite types there is a significant departure in $\delta^{18}O$ values from the postulated range of dolomite that was precipitated in equilibrium with seawater of respective ages. This can be related to increasing temperature during burial as well as modification of pore fluid isotopic composition, in SD this negative shift can be related to the effects of hydrothermal fluid interaction with host rocks; perhaps limited in extent and distribution. Salinity and Th measurements in the Silurian also show a distinctive pattern with higher and salinity temperatures of formation comparing to other rock units, which suggest the effect of evaporite dissolution and migration of fluids in these rocks.

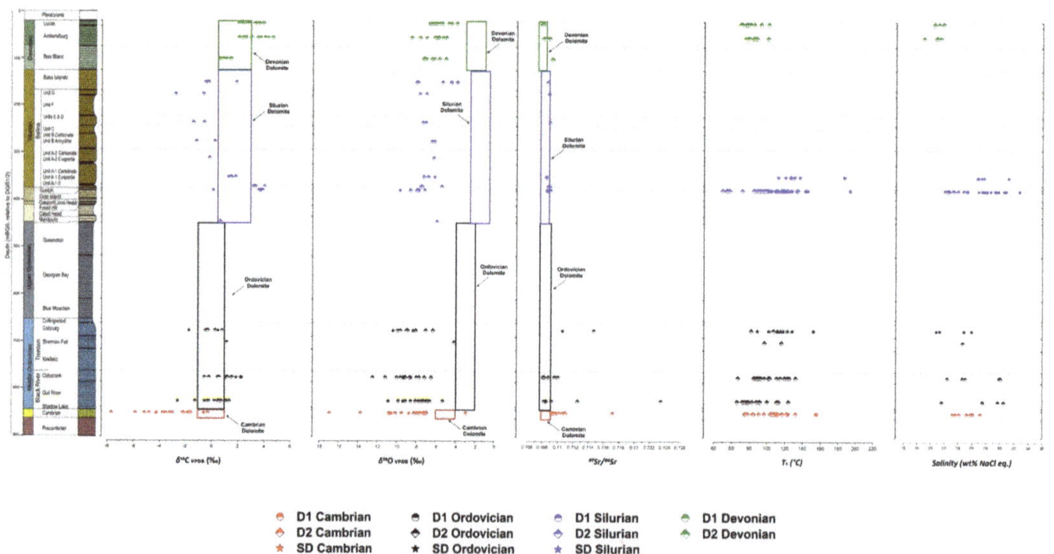

Figure 24. Stratigraphy vs. isotopic and microthermometric data for the investigate age groups.

The conceptual model for fluid flow mechanism in the Huron Domain shown in Figure 25 illustrates the concept of fluid compartmentalization in different stratigraphic rock units suggested in this study. The current burial depth is shown also in the model based on various studies [80,111,120]. As mentioned earlier this region of the Michigan

Basin has experienced multiple stages of tectonic evolution, both passive and active tectonic events which affected fluid composition and migration pathways (e.g., [2,5,32,37]). Migration of warm, basinal brines from the center of Michigan Basin towards the margin [32] transported dolomitizing fluids and other solutes. The source of heat of these fluids could be related to the reactivation of the buried mid-continental rifts during late Devonian to Mississippian time [121]. A limited influx of hydrothermal fluids moved upwards from the Precambrian basement via fractures and faults [2,47] results in minor precipitation of saddle dolomite in fractures and vugs and altered earlier formation of matrix dolomite. This likely occurred during Taconian and Acadian Orogenies. Interaction of connate fluids of marine parentage with warm, basinal brines controlled the formation of dolomite and calcite cement at a shallow to intermediate burial depth. We believe that diagenesis is horizontally (stratabound) uniform across the Huron domain, displaying few signs of significant vertical connectivity beyond formation tops, except in areas with local heterogeneity (faults; [122]).

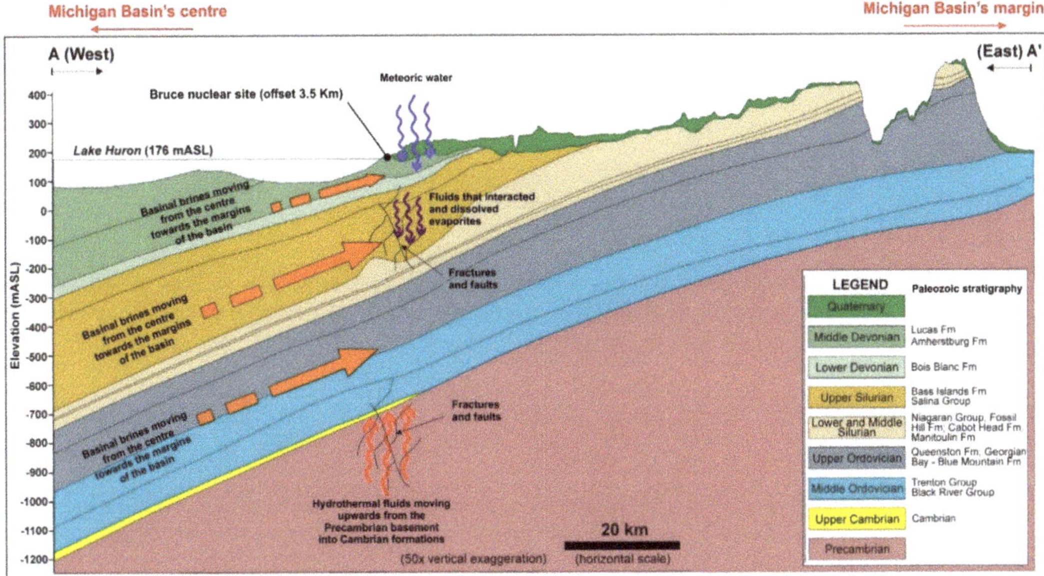

Figure 25. Conceptual model for fluid flow in the study area illustrating fluid compartmentalization in different stratigraphic rock units and migration pathways. These fluid pathways vary with time and effect of tectonic setting (stratigraphic setting is modified from [123]).

Downward migration of saline fluids that originated from the dissolution of Silurian evaporites and possible interaction with basinal brines moving up-dip for the center of the basin was responsible for the modification of diagenetic phases [33]. In the Devonian hydrogeochemical system, the interaction of basinal brines with possible infiltration of younger fluids, perhaps of meteoric water origin, resulted in the formation of late dolomite and calcite cement (e.g., [55,111,112]). This probably occurred during the Alleghenian Orogeny (e.g., [3]). Sutcliff et al. [124] in their study of late calcite cement recovered from sub-horizontal fractures near the base of the Silurian sequence dated show U–Pb ages of 318 ± 10 Ma determined by LA-ICP-MS and 313 ± 1 Ma by ID-TIMS. Hence, these authors also suggested that the fluid flow of hydrothermal brines was influenced by the Alleghanian Orogeny which caused the migration of hydrothermal fluids westwards from high-level Alleghanian plutons through the basal Cambrian rocks.

6. Conclusions

Based on petrographic, geochemical, and fluid inclusion data presented, the following conclusions have been drawn with regards to the diagenetic evolution of the Huron Domain of southern Ontario:

- Dolomitization of Cambrian to Devonian carbonates occurred through multiple events, with the formation of the earliest replacive dolomite phase (D1) occurring at shallow burial depths. This was followed by the precipitation of a later replacive dolomite phase (D2) and the formation of dolomite cement (SD) occluding vugs and fractures, which were formed at an intermediate burial depth. Evaporite and late blocky calcite cement followed the formation of these dolomite phases.
- These types of dolomite types display unique geochemical and microthermometric signatures suggesting the flow of diagenetic fluids that were driven by tectonic processes over the history of the Michigan Basin. The diagenetic fluid composition was dominated initially by seawater and later modified upon burial and interaction with basinal brines. For Cambrian and Ordovician carbonates two possibly isolated diagenetic fluid systems affected them. These are: (i) an earlier fluid system that is characterized by a pronounced negative shift in oxygen and carbon isotopic composition, more radiogenic ratios, warm and saline signature with higher average $\sum$REE, negative La anomaly and positive Ce anomaly; and (ii) a later Ordovician system, characterized by less negative shifts in oxygen and carbon isotopes, comparable homogenization temperature, hypersaline, a less radiogenic fluid system, less negative La anomaly and positive and negative Ce anomaly, but also higher average $\sum$REE compared to warm water marine brachiopods.
- There is a significant overlap of $\delta^{13}C$ and $\delta^{18}O$ values and $^{87}Sr/^{86}Sr$ ratios in Silurian and Devonian carbonates reflecting both a modified marine source of fluids as well recrystallization of earlier formed dolomite. High homogenization temperatures and salinity in saddle dolomite (SD) in the Silurian sequences and its REE shale-normalized patterns suggest that in both age groups the diagenetic fluids were originally of coeval seawater composition and were subsequently modified via water-rock interaction. The different evolution of the diagenetic fluids is more prominent in REE_{SN} patterns from Silurian samples and it is possibly related to brines, which were modified by the dissolution of Silurian evaporites from the Salina series.

Author Contributions: Conceptualization, I.S.A.-A.; writing—original draft preparation, I.S.A.-A.; supervision, I.S.A.-A.; writing, review and editing, R.C.; writing, review and editing, M.T. All authors have read and agreed to the published version of the manuscript.

Funding: This research is funded by Nuclear Waste Management Organization (NWMO) contract #816150 to ISA.

Institutional Review Board Statement: Not applicable.

Informed Consent Statement: Not applicable.

Data Availability Statement: The data will be available with the authors.

Acknowledgments: I.S.A.-A. and M.T. acknowledge the help from M. Price during isotope analysis. The critical review by M. Hobbs is greatly appreciated. Thanks to two anonymous reviewers for their constructive comments.

Conflicts of Interest: The authors declare no conflict of interest.

References

1. Al-Aasm, I.S. *Dolomitization: Cambrian and Ordovician Formations in the Huron Domain*; Report NWMO-TR-2016-05; Nuclear Waste Management Organization: Toronto, ON, Canada, 2016.
2. Al-Aasm, I.S.; Crowe, R. Fluid compartmentalization and dolomitization in the Cambrian and Ordovician successions of the Huron Domain, Michigan Basin. *Mar. Pet. Geol.* **2018**, *92*, 160–178. [CrossRef]

3. Tortola, M.; Al-Aasm, I.S.; Crowe, R. Diagenetic Pore Fluid Evolution and Dolomitization of the Silurian and Devonian Carbonates, Huron Domain of Southwestern Ontario: Petrographic, Geochemical and Fluid Inclusion Evidence. *Minerals* **2020**, *10*, 140. [CrossRef]
4. Muchez, P.; Slobodnik, M.; Sintubin, M.; Viaene, W.; Keppens, E. Origin and migration of palaeofluids in the Lower Carboniferous of southern and eastern Belgum. *Zent. Geol. Palaeontol.* **1997**, *11 Pt 1*, 1107–1112.
5. Bethke, C.M.; Marshak, S. Brine migrations across North America—The plate tectonics of groundwater. *Annu. Rev. Earth Planet. Sci.* **1990**, *18*, 287–315. [CrossRef]
6. Kharaka, Y.K.; Hanor, J.S. Deep fluids in the continents: I. sedimentary basins. In *Treatise on Geochemistry*; Drever, J.I., Holland, H.D., Turekian, K.K., Eds.; Elsevier: Amsterdam, The Netherlands, 2004; Volume 5, pp. 499–540. [CrossRef]
7. Morad, S.; Al-Ramadan, K.; Ketzer, J.M.; De Ros, L.F. The impact of diagenesis on the heterogeneity of sandstone reservoirs: A review of the role of depositional facies and sequence stratigraphy. *AAPG Bull.* **2010**, *94*, 1267–1309. [CrossRef]
8. Durocher, S.; Al-Aasm, I.S. Dolomitization and neomorphism of Mississippian (Visean) upper Debolt Formation, Blueberry Field, NE British Columbia, Canada: Geologic, petrologic and chemical evidence. *Am. Assoc. Pet. Geol. Bull.* **1997**, *81*, 954–977.
9. Lonnee, J.; Al-Aasm, I.S. Dolomitization and fluid evolution in the Middle Devonian Sulphur. Point Formation, Rainbow South Field, Alberta: Petrographic and geochemical evidence. *Bull. Can. Soc. Pet. Geol.* **2000**, *90*, 262–283. [CrossRef]
10. Machel, H.G.; Lonnee, J. Hydrothermal dolomite—A product of poor definition and imagination. *Sediment. Geol.* **2002**, *152*, 163–171. [CrossRef]
11. Cioppa, M.T.; Al-Aasm, I.S.; Symons, D.T.A.; Gillen, K.P. Dating penecontemporaneous dolomitization in carbonate reservoirs: Paleomagnetic, petrographic, and geochemical constraints. *AAPG Bull.* **2003**, *87*, 71–88.
12. Wendte, J.; Chi, G.; Al-Aasm, I.; Sargent, D. Fault/fracture controlled hydrothermal dolomitization and associated diagenesis of the Upper Devonian Jean Marie Member (Redknife Formation) in the July Lake area of northeastern British Columbia. *Bull. Can. Pet. Geol.* **2009**, *57*, 275–322. [CrossRef]
13. Al-Aasm, I.S.; Mrad, C.; Packard, J. Fluid compartmentalization of Devonian and Mississippian dolostones, Western Canada Sedimentary Basin: Petrologic and geochemical evidence from fracture mineralization. *Can. J. Earth Sci.* **2019**, *56*, 265–305. [CrossRef]
14. Dong, S.; Chen, D.; Zhou, X.; Qian, Y.; Tian, M.; Qing, H. Tectonically driven dolomitization of Cambrian to Lower Ordovician carbonates of the Quruqtagh area, north-eastern flank of Tarim Basin, north-west China. *Sedimentology* **2016**, *64*, 1079–1106. [CrossRef]
15. Land, L.S. The origin of massive dolomite. *J. Geol. Educ.* **1985**, *33*, 112–125. [CrossRef]
16. Morrow, D.W. Dolomite—Part 2: Dolomitization Models and Ancient Dolostones. In *Diagenesis*; McIlreath, I.A., Morrow, D.W., Eds.; Geoscience Canada Series 4; Geological Association of Canada: St John's, NL, Canada, 1990; pp. 125–139.
17. Packard, J.J.; Pellegrin, G.J.; Al-Aasm, I.S.; Samson, I.; Gagnon, J. Diagenesis and dolomitization associated with hydrothermal karst in Famennian upper Wabamun ramp sediments, northwestern Alberta. The development of porosity in carbonate reservoirs. In *Canadian Society of Petroleum Geologists Short Course*; The Development of Porosity in Carbonate Reservoirs: Calgary, AB, Canada, 1990; Volume 9-1.
18. Sperber, C.M.; Wilkinson, B.H.; Peacor, D.R. Rock composition, dolomite stoichiometry, and rock/water reactions in dolomitic carbonate rocks. *J. Geol.* **1984**, *92*, 609–622. [CrossRef]
19. Gregg, J.M.; Shelton, K.L. Dolomitization and dolomite Neomorphism in the back-reef facies of the Bonneterre and Davis formations (Cambrian), southwestern Missouri. *J. Sediment. Res.* **1990**, *60*, 549–562.
20. Gao, G.; Land, L.S. Early Ordovician cool creel dolomite, middle Arbuckle group, silick Hills, SW Oklahoma, USA: Origin and modification. *J. Sediment. Res.* **1991**, *61*, 161–173.
21. Mazzullo, S.J. Geochemical and neomorphic alteration of dolomite: A review. *Carbonate Evaporites* **1992**, *7*, 21–37. [CrossRef]
22. Kupecz, J.A.; Land, L.S. Progressive recrystallization and stabilization of early-stage dolomite: Lower Ordovician Ellenburger Group, west Texas. In *Dolomites. A Volume in Honor of Dolomieu*; Purser, B., Tucker, M., Zenger, D., Eds.; International Association of Sedimentologists: Gent, Belgium, 1994; pp. 155–166.
23. Al-Aasm, I.S. Chemical and isotopic constraints for recrystallization of sedimentary dolomites from the Western Canada Sedimentary basin. *Aquat. Geochem.* **2000**, *6*, 227–248. [CrossRef]
24. Oliver, J. Fluids expelled tectonically from orogenic belts: Their role in hydrocarbon migration and other geologic phenomena. *Geology* **1986**, *14*, 99–102. [CrossRef]
25. Nesbitt, B.E.; Muehlenbachs, K. Paleohydrogeology of the Canadian Rockies and origins of brines, Pb-Zn deposits and dolomitization in the Western Canada Sedimentary Basin. *Geology* **1994**, *22*, 243–246. [CrossRef]
26. Al-Aasm, I.S. Origin and characterization of hydrothermal dolomite in the Western Canada Sedimentary Basin. *J. Geochem. Explor.* **2003**, *78*, 9–15. [CrossRef]
27. Kareem, H.K.; Al-Aasm, I.S.; Mansurbeg, H. Stucturally-controlled hydrothermal fluid flow in an extensional tectonic regime: A case study of Cretaceous Qamchuqa Formation, Zagros Basin, Kurdistan Iraq. *Sediment. Geol.* **2019**, *386*, 52–78. [CrossRef]
28. Tucker, M.E.; Wright, V.P. *Carbonate Sedimentology*; Blackwell Scientific Publications: Oxford, UK, 1990; 482p.
29. Warren, J. Dolomite: Occurrence, evolution, and economically important associations. *Earth Sci. Rev.* **2000**, *52*, 1–81. [CrossRef]

30. Machel, H.G. Concepts and models of dolomitization: A critical reappraisal. In *The Geometry and Petrogenesis of Dolomite Hydrocarbon Reservoirs*; Braithwaite, C.J.R., Rizzi, G., Drake, G., Eds.; Geological Society Special Publication: London, UK, 2004; Volume 235, pp. 7–63.
31. Davies, G.R.; Smith, L.B., Jr. Structurally controlled hydrothermal dolomite reservoir facies: An overview. *Am. Assoc. Pet. Geol.* **2006**, *90*, 1641–1690. [CrossRef]
32. Haeri-Ardakani, O.; Al-Aasm, I.S.; Coniglio, M. Fracture mineralization and fluid flow evolution: An example from Ordovician-Devonian carbonates, southwestern Ontario, Canada. *Geofluids* **2013**, *13*, 1–20. [CrossRef]
33. Haeri-Ardakani, O.; Al-Aasm, I.S.; Coniglio, M. Petrologic and geochemical attributes of fracture-related dolomitization in Ordovician carbonates and their spatial distribution in southwestern Ontario, Canada. *Mar. Pet. Geol.* **2013**, *43*, 409–422. [CrossRef]
34. Carter, T.R.; Fortner, L.D.; Russell, H.J.A.; Skuce, M.E.; Longstaffe, F.J.; Sun, S. A Hydrostratigrapic Framework for the Paleozoic Bedrock of Southern Ontario. *Geosci. Can.* **2021**, *48*, 23–58. [CrossRef]
35. Ontario Geological Survey. *Bedrock Geology of Ontario, Southern Sheet, Map 2544, Scale 1:1,000,000*; Ontario Geological Survey: Sudbury, ON, Canada, 1991.
36. Ontario Power Generation's (OPG). *Deep Geologic Repository for Low and Intermediate Level Waste: Preliminary Safety Report*; Program Document n 00216-SR-01320-00001; OPG: Thunder Bay, ON, Canada, 2011.
37. Sanford, B.V.; Thompson, F.J.; McFall, G.H. Plate tectonics—A possible controlling mechanism in the development of hydrocarbon traps in southwestern Ontario. *Bull. Can. Pet. Geol.* **1985**, *33*, 52–71.
38. Cruden, A. *Outcrop Fracture Mapping*; NMWO DGR-TR-2011-43; Nuclear Waste Management Organization: Toronto, ON, Canada, 2011.
39. Hamilton, D.G.; Coniglio, M. Reservoir Development in the Middle Devonian of southwestern Ontario. In Proceedings of the Ontario Petroleum Institute Thirty-First Annual Conference, Niagara Falls, ON, Canada, 28–30 October 1992; Technical Paper No. 9. Volume 32. 20p.
40. Armstrong, D.K.; Carter, T.R. *The Subsurface Paleozoic Stratigraphy of Southern Ontario*; Ontario Geological Survey: Sudbury, ON, Canada, 2010; Special Volume 7, 301p.
41. Klein, G.V.; Hsui, A.T. Origin of Cratonic Basins. *Geology* **1987**, *15*, 1094–1098. [CrossRef]
42. Sleep, N.H. Thermal effects of the formation of Atlantic continental margins by continental break up. *Geophys. J. R. Astron. Soc.* **1971**, *24*, 325–350. [CrossRef]
43. Sleep, N.H.; Snell, N.S. Thermal contraction and flexure of mid-continent and Atlantic marginal basins. *Geophys. J. R. Astron. Soc.* **1976**, *45*, 125–154. [CrossRef]
44. Nunn, J.A.; Sleep, N.H.; Moore, W.E. Thermal subsidence and generation of hydrocarbon in Michigan Basin. *AAPG Bull.* **1984**, *68*, 296–315.
45. Howell, P.D.; van der Pluijm, B.A. Early history of the Michigan basin: Subsidence and Appalachian tectonics. *Geology* **1990**, *18*, 1195–1198. [CrossRef]
46. Howell, P.D.; van der Pluijm, B.A. Structural sequences and styles of subsidence in the Michigan basin. *Geol. Soc. Am. Bull.* **1999**, *111*, 974–991. [CrossRef]
47. AECOM; Itasca Consulting Canada. *Regional Geology—Southern Ontario*; Report NWMO DGR-TR-2011-15 R000; Nuclear Waste Management Organization: Toronto, ON, Canada, 2011.
48. Coakley, B.J.; Nadon, C.; Wang, H.F. Spatial variations in tectonic subsidence during Tippecanoe I in the Michigan Basin. *Basin Res.* **1994**, *6*, 131–140. [CrossRef]
49. Coakley, B.; Gurnis, M. Far-field tilting of Laurentia during the Ordovician and constraints on the evolution of a slab under an ancient continent. *J. Geophys. Res. Solid Earth* **1995**, *100*, 6313–6327. [CrossRef]
50. Johnson, M.D.; Armstrong, D.K.; Sanford, V.V.; Telford, P.G.; Rutka, M.A. Paleozoic and Mesozoic Geology of Ontario. In *Geology of Ontario*; Ontario Geologic Survey: Peterborough, ON, Canada, 1992; Special Volume 4, Part 2, pp. 907–1008.
51. Armstrong, D.K.; Carter, T.R. An updated guide to the subsurface Paleozoic stratigraphy of Southern Ontario. *Ont. Geol. Surv.* **2006**, *6191*, 214.
52. Beaumont, C.; Quinlan, G.; Hamilton, J. Orogeny and stratigraphy: Numerical models of the Paleozoic in the eastern interior of North America. *Tectonics* **1988**, *7*, 389–416. [CrossRef]
53. Nuclear Waste Management Organization. Deep Geologic Repository, Geoscientific Site Characterization. 2011. Available online: http//www.nmwo.ca/dggeoscientifcsitecharaterization (accessed on 29 August 2021).
54. Nuclear Waste Management Organization. Geosynthesis: Nuclear Waste Management Organization. Report NWMO DGR-TR-2011-11. Toronto, Canada. 2011. Available online: http//www.nmwo.ca/uploads.DGR%20PDF/Geo/Geosynthesis/pdf (accessed on 29 August 2021).
55. Middleton, K.; Coniglio, M.; Sherlock, R.; Frape, S.K. Dolomitization of Middle Ordovician carbonate reservoirs, southwestern Ontario. *Bull. Can. Pet. Geol.* **1993**, *41*, 150–163.
56. Intera Engineering. *Descriptive Geosphere Site Model*; Report NWMO DGR-TR-2011-24 R000; Nuclear Waste Management Organization: Toronto, ON, Canada, 2011.
57. Van der Voo, R. Paleozoic paleogeography of North America, Gondwana, and intervening displaced terranes: Comparison of paleomagnetism with paleoclimatology and biogeographical patterns. *Geol. Soc. Am. Bull.* **1988**, *100*, 311–324. [CrossRef]

58. Sonnenfeld, P.; Al-Aasm, I.S. The Salina evaporites in the Michigan Basin. In *Early Sedimentary Evolution of the Michigan Basin*; Special Paper, 256; Catacosinos, P.A., Daniels, P.A., Jr., Eds.; Geological Society of America: Boulder, CO, USA, 1991; pp. 139–153.
59. Birchard, M.C.; Rutka, M.A.; Brunton, F.R. *Lithofacies and Geochemistry of the Lucas Formation in the Subsurface of Southwestern Ontario: A High Purity Limestone and Potential High Purity Dolostone Resource*; Open File Report 6137; Ontario Geological Survey: Sudbury, ON, Canada, 2004; 57p.
60. Al-Aasm, I.S.; Taylor, B.; South, B. Stable isotope analysis of multiple carbonate samples using selective acid extraction. *Chem. Geol.* **1990**, *80*, 119–125. [CrossRef]
61. Chi, G.; Ni, P. Equations for calculation of NaCl/(NaCl+$CaCl_2$) ratios and salinities from hydrohalite-melting and ice-melting temperatures in the H_2O-NaCl-$CaCl_2$ system. *Acta Petrol. Sin.* **2007**, *23*, 33–37.
62. Steele-MacInnis, M.; Bondnar, R.J.; Naden, J. Numerical model to determine the composition of H2O-NaCl-CaCl2 fluid inclusions based on microthermometirc and microanalytical data. *Geochem. Cosmochim. Acta* **2011**, *75*, 21–41. [CrossRef]
63. Lawler, J.P.; Crawford, M.L. Stretching of fluid inclusions resulting from a low temperature microthermometric technique. *Econ. Geol.* **1983**, *78*, 527–529. [CrossRef]
64. Jenner, G.A.; Longerich, H.P.; Jackson, S.E.; Fryer, B.J. ICP-MS—A powerful tool for high-precision trace-element analysis in Earth sciences: Evidence from analysis of selected USGS refence samples. *Chem. Geol.* **1990**, *83*, 133–148. [CrossRef]
65. Bau, M.; Dulski, P. Distribution of yttrium and rare-earth elements in the Penge and Kuruman iron-formations, Transvaal Supergroup, South Africa. *Precambrian Res.* **1996**, *79*, 37–55. [CrossRef]
66. Kucera, J.; Cempirek, J.; Dolnicek, Z.; Muchez, P.; Prochaska, W. Rare earth elements and yttrium geochemistry of dolomite from post-Variscan vein-type mineralization of the Nizky Jesenik and Upper Silesian Basins, Czech Republic. *J. Geochem. Explor.* **2009**, *103*, 69–79. [CrossRef]
67. Felitsyn, S.; Morad, S. REE patterns in latest Neoproterozoic–early Cambrian phosphate concretions and associated organic matter. *Geochem. Geol.* **2002**, *187*, 257–265. [CrossRef]
68. Sibley, D.F.; Gregg, J.M. Classification of dolomite rock textures. *J. Sediment. Petrol.* **1987**, *57*, 967–975.
69. Veizer, J.; Ala, D.; Azmy, K.; Bruckschen, P.; Buhl, D.; Bruhn, F.; Carden, G.A.F.; Diener, A.; Ebneth, S.; Godderis, Y.; et al. $^{87}Sr/^{86}Sr$ and $\delta^{13}C$ and $\delta^{18}O$ evolution of Phanerozoic seawater. *Chem. Geol.* **1999**, *161*, 59–88. [CrossRef]
70. Azmy, K.; Brand, U.; Sylvester, P.; Gleeson, S.A.; Logan, A.; Bitner, M.A. Biogenic and abiogenic low-Mg calcite (bLMC and aLMC): Evaluation of seawater-REE composition, water masses and carbonate diagenesis. *Chem. Geol.* **2011**, *280*, 180–190. [CrossRef]
71. Goldstein, R.H.; Reynolds, T.J. *Systematics of Fluid Inclusions in Diagenetic Minerals. Society of Economic Paleontologists and Mineralogists*; Society for Sedimentary Geology: Tulsa, OK, USA, 1994; 199p.
72. Morad, S.; Al-Aasm, I.S.; Nader, F.H.; Ceriani, A.; Gasparrani, M. Impact of diagenesis on the spatial and temporal of reservoir quality in C and D Members of the Arab Formation (Jurassic), offshore Abu Dhabi oilfield. *GeoArabia* **2012**, *17*, 17–56.
73. Harper, D.A.; Longstaffe, F.J.; Wadleigh, M.A.; McNutt, R.H. Secondary K-feldspar at the Precambrian-Paleozoic unconformity, southwestern Ontario. *Can. J. Earth Sci.* **1995**, *32*, 1432–1450. [CrossRef]
74. Ziegler, K.; Longstaffe, F.J. Clay mineral anthogenesis along a mid-continental scale fluid conduit in Palaeozoic sedimentary rocks from southern Ontario, Canada. *Clay Miner.* **2001**, *35*, 239–260. [CrossRef]
75. Paganoni, M.; Al Harthi, A.; Morad, D.; Morad, S.; Ceriani, A.; Mansurbeg, H. Impact of stylolitization on diagenesis of a Lower Cretaceous carbonate reservoir from a giant oilfield, Abu Dhabi, United Arab Emirates. *Sediment. Geol.* **2016**, *335*, 70–92. [CrossRef]
76. Sanford, B.V. Devonian of Ontario and Michigan. *Int. Symp. Devonian Syst.* **1968**, *1*, 973–999.
77. Armstrong, D.K.; Goodman, W.R. Stratigraphy and Depositional Environments of Niagaran Carbonates, Bruce Peninsula, Ontario. In *Proceedings of the Spring Eastern Section Meeting, University Park, PA, USA, 7–8 April 1990*; Field Trip No. 4 Guidebook; American Association of Petroleum Geologists: London, ON, Canada; Ontario Petroleum Institute: Tulsa, OK, USA, 1990.
78. James, N.P.; Jones, B. *Origin of Carbonate Sedimentary Rocks*; John Wiley & Sons: West Sussex, UK, 2016; p. 446.
79. Cercone, K.R. Evaporative sea-level drawdown in the Silurian Michigan Basin. *Geology* **1988**, *16*, 109–130. [CrossRef]
80. Zheng, Q. Carbonate Diagenesis and Porosity Evolution in the Guelph Formation, Southwestern Ontario. Ph.D. Thesis, University of Waterloo, Waterloo, ON, Canada, 1999.
81. Swart, P.K. The geochemistry of carbonate diagenesis: The past, present and future. *Sedimentology* **2015**, *62*, 1233–1304. [CrossRef]
82. Frank, T.D.; Lohmann, K.C. Chronostratigraphic significance of cathodoluminescence zoning in syntaxial cement: Mississippian Lake Valley Formation, New Mexico. *Sediment. Geol.* **1996**, *105*, 29–50. [CrossRef]
83. McNutt, R.H.; Frape, S.K.; Dollar, P. A strontium, oxygen and hydrogen isotopic composition of brines, Michigan and Appalachian Basins, Ontario and Michigan. *Appl. Geochem.* **1987**, *2*, 495–505. [CrossRef]
84. Baldermann, A.; Deditius, A.; Dietzel, M.; Fichtner, V.; Fischer, C.; Hippler, D.; Leis, A.; Baldermann, C.; Mavromatis, V.; Stickler, C.P.; et al. The role of bacterial sulfate reduction during dolomite precipitation: Implications from Upper Jurassic platform carbonates. *Chem. Geol.* **2015**, *412*, 1–14. [CrossRef]
85. Azomani, E.; Azmy, K.; Blemey, N.; Brand, U.; Al-Aasm, I.S. Dolomitization of the Catoche Formation in Daniel's Harbour, Northern Peninsula, western Newfoundland. *Mar. Pet. Geol.* **2013**, *40*, 99–114. [CrossRef]
86. Al-Aasm, I.; Veizer, J. Diagenetic stabilization of aragonite and low-Mg calcite-1. Trace elements in rudists. *J. Sed. Petrol.* **1986**, *56*, 138–152. [CrossRef]

87. Land, L.S. The isotopic and trace element geochemistry of dolomite: The state of the art. In *Concepts and Models of Dolomitization*; Zenger, D.H., Dunham, J.B., Ethington, R.L., Eds.; SEPM Special Publication: Tulsa, OK, USA, 1980; Volume 28, pp. 87–110.
88. Nothdurft, L.; Webb, G.; Kamber, B. Rare earth element geochemistry of Late Devonian reefal carbonates, Canning Basin, Western Australia: Confirmation of a seawater REE proxy in ancient limestones. *Geochim. Cosmochim. Acta* **2004**, *68*, 263–283. [CrossRef]
89. Webb, G.E.; Nothdurft, L.; Kamber, B.; Kloprogge, T.; Zhao, J.-X. Rare earth element geochemistry of scleractinian coral skeleton during meteoric diagenesis: A before and-after sequence through neomorphism of aragonite to calcite. *Sedimentology* **2009**, *56*, 1433–1463. [CrossRef]
90. Özyurt, M.; Kirmachi, M.Z.; Al-Aasm, I.; Hollis, C.; Tasli, K.; Kandemir, R. REE Characteristics of Lower Cretaceous Limestone Succession in Gümüshane, NE Turkey: Implications for Ocean Paleoredox Conditions and Diagenetic Alteration. *Minerals* **2020**, *10*, 683. [CrossRef]
91. Sholkovitz, E.; Shen, G.T. The incorporation of rare earth elements in modern coral. *Geochim. Cosmochim. Acta* **1995**, *59*, 2749–2756. [CrossRef]
92. Allwood, A.C.; Kamber, B.S.; Walter, M.R.; Burch, I.W.; Kanik, I. Trace elements record depositional history of an Early Archean stromatolitic carbonate platform. *Chem. Geol.* **2010**, *270*, 148–163. [CrossRef]
93. Shields, G.; Stille, P. Diagenetic constraints on the use of cerium anomalies as paleoseawater redox proxies: An isotopic and REE study of Cambrian phosphorites. *Chem. Geol.* **2001**, *175*, 29–48. [CrossRef]
94. Sverjensky, D.A. Europium equilibria in aqueous solution. *Earth Planet. Sci. Lett.* **1984**, *67*, 70–78. [CrossRef]
95. Zaky, A.H.; Brand, U.; Azmy, K. A new sample processing protocol for procuring seawater REE signatures in biogenic and abiogenic carbonates. *Chem. Geol.* **2015**, *416*, 36–50. [CrossRef]
96. Caetano-Filho, S.; Paula-Santos, G.M.; Dias-Brito, D. Carbonate REE + Y signatures from the restricted early marine phase of South Atlantic Ocean (late Aptian—Albian): The influence of early anoxic diagenesis on shale-normalized REE + Y patterns of ancient carbonate rocks. *Palaeogeogr. Palaeoclimatol. Palaeoecol.* **2008**, *500*, 69–83. [CrossRef]
97. Hood, A.S.; Planavsky, N.J.; Wallace, M.W.; Wang, X. The effects of diagenesis on geochemical paleoredox proxies in sedimentary carbonates. *Geochim. Cosmochim. Acta* **2018**, *232*, 265. [CrossRef]
98. Worthington, S.R.H. *Karst Assessment*; Report NWMO DGR-TR-2011-22; Nuclear Waste Management Organization: Toronto, ON, Canada, 2011.
99. Banner, J.L.; Hanson, G.N. Calculation of simultaneous isotopic and trace element variations during water-rock interaction with applications to carbonate diagenesis. *Geochim. Cosmochim. Acta* **1990**, *54*, 3123–3137. [CrossRef]
100. Özyurt, M.; Kirmachi, M.Z.; Al-Aasm, I.S. Geochemical characteristics of Late Jurassic-Early Cretaceous Berdiga carbonates in Hazine Mağara, Gümüşhane (NE Turkey): Implications for dolomitization and recrystallization. *Can. J. Earth Sci.* **2019**, *56*, 306–320.
101. Luczaj, J.A. Evidence against the Dorag (mixing-zone) model for dolomitization along the Wisconsin arch-A case for hydrothermal diagenesis. *Am. Assoc. Pet. Geol. Bull.* **2006**, *90*, 1719–1738. [CrossRef]
102. Barnes, D.A.; Parris, T.M.; Grammer, G.M. Hydrothermal Dolomitization of Fluid Reservoirs in the Michigan Basin, USA. Search and Discovery Article #50087. In Proceedings of the AAPG Annual Convention, San Antonio, TX, USA, 20–23 April 2008.
103. Simo, J.A.; Johnson, C.M.; Vandrey, M.R.; Brown, P.E.; Castro Giovanni, E.; Drzewiecki, P.E.; Valley, J.W.; Boyer, J. Burial dolomitization of the Middle Ordovician Glenwood Formation by evaporitic brines, Michigan Basin. In *Dolomites—A Volume in Honour of Dolomieu*; Purser, B., Tucker, M., Zenger, D., Eds.; The International Association of Sedimentologists, Blacwell Scientific Publications: Hoboken, NJ, USA, 1994; pp. 169–186.
104. Braithwaite, C.J.R.; Rizzi, G.; Darke, G. The geometry and petrogenesis of dolomite hydrocarbon reservoirs: Introduction. *Geol. Soc. Lond. Spec. Publ.* **2004**, *235*, 1–6. [CrossRef]
105. Machel, H.G. Recrystallization versus neomorphism, and the concept of "significant recrystallization" in dolomite research. *Sediment. Geol.* **1997**, *113*, 161–168. [CrossRef]
106. Kaczmarek, S.E.; Sibley, D.F. Direct physical evidence of dolomite recrystallization. *Sedimentology* **2014**, *61*, 1862–1882. [CrossRef]
107. Adam, J.; Al-Aasm, I.S. Petrologic and geochemical attributes of calcite cementation, dolomitization and dolomite recrystallization: An example from the Mississippian Pekisko Formation, west-central Alberta. *Bull. Can. Pet. Geol.* **2017**, *65*, 235–261. [CrossRef]
108. Hardie, L.A. Perspectives—dolomitization: A critical view of some current views. *J. Sediment. Petrol.* **1987**, *57*, 166–183. [CrossRef]
109. Al-Aasm, I.S.; Clarke, J.D. The effect of hydrothermal fluid flow on early diagenetic dolomitization: An example from the Devonian Slave Point Formation, northwest Alberta, Canada. In *Deformation, Fluid Flow, and Reservoir Appraisal in Foreland Fold and Thrust Belts: AAPG Hedberg Series*; Swennen, R., Roure, F., Granath, J.W., Eds.; American Association of Petroleum Geologists: Tulsa, OK, USA, 2004; pp. 297–316.
110. Malone, M.J.; Baker, P.A.; Burns, S.J. Recrystallization of dolomite: An experimental study from 50–200 °C. *Geochim. Cosmochim. Acta* **1996**, *60*, 2189–2207. [CrossRef]
111. Coniglio, M.; William-Jones, A.E. Diagenesis of Ordovician carbonates from the northeast Michigan Basin, Manitoulin Island area, Ontario: Evidence from petrography, stable isotopes and fluid inclusions. *Sedimentology* **1992**, *39*, 813–836. [CrossRef]
112. Coniglio, M.; Sherlock, R.; Williams-Jones, A.E.; Middleton, K.; Frape, S.K. Burial and hydrothermal diagenesis of Ordovician carbonates from the Michigan Basin, Ontario, Canada. *Int. Assoc. Sedimentol.* **1994**, *21*, 231–254.
113. Yoo, C.M.; Gregg, J.M.; Shelton, K.L. Dolomitization and dolomite neomorphism: Trenton and Black River limestone (Middle Ordovician) Northern Indiana, U.S.A. *J. Sediment. Res.* **2000**, *70*, 265–274. [CrossRef]

114. Hobbs, M.Y.; Frape, S.K.; Shouakar-Stash, O.; Kennell, L.R. *Regional Hydrochemistry-Southern Ontario*; Report NMWO-DGR-TR-2011-12; Nuclear Waste Management Organization: Toronto, ON, Canada, 2011.
115. Petts, D.; Saso, J.; Diamond, L.; Aschwanden, L.; Al, T.; Jensen, M. The source and evolution of paleofluids responsible for secondary minerals in low-permeability Ordovician limestones of the Michigan Basin. *Appl. Geochem.* **2017**, *86*, 121–137. [CrossRef]
116. Bouchard, L.; Veizer, J.; Kennell-Morrison, L.; Jensen, M.; Raven, K.G.; Clark, I.D. Origin and ^{87}Rb–^{87}Sr age of porewaters in low permeability Ordovician sediments on the eastern flank of the Michigan Basin, Tiverton, Ontario, Canada. *Can. J. Earth Sci.* **2019**, *56*, 201–208. [CrossRef]
117. Nagia, N.G.; Hu, M.; Gao, D.; Hu, Z.; Sun, C.-Y. Application of stable strontium isotope geochemistry and fluid inclusion microthermometry to studies of dolomitization of the deeply buried Cambrian carbonate successions in west-central Tarim Basin, NW China. *J. Earth Sci.* **2019**, *30*, 176–193. [CrossRef]
118. Land, L.S. The application of stable isotopes to studies of the origin of dolomite and to problems of diagenesis of clastic sediments. In *Stable Isotopes in Sedimentary Geology*; Arthur, M.A., Anderson, T.F., Kaplan, I.R., Veizer, J., Land, L.S., Eds.; SEPM Society for Sedimentary Geology: Tulsa, OK, USA, 1983.
119. Friedman, I.; O'Neil, J.R. Compilation of stable isotope fractionation factors of geochemical interest. In *Data of Geochemistry*, 6th ed.; Chapter, K.K., Ed.; U.S. Government Printing Office: Washington, DC, USA, 1977.
120. Wang, H.F.; Crowley, K.D.; Nadon, G.C. Thermal history of the Michigan Basin from Apatite fission-track analysis and vitrinite reflectance. *Am. Assoc. Petrol. Geol. Mem.* **1994**, *61*, 167–177.
121. Ma, L.; Castro, M.C.; Hall, C.M. Atmospheric noble gas signatures in deep Michigan Basin brines as indicators of a past thermal event. *Earth Planet Sci. Lett.* **2009**, *277*, 137–147. [CrossRef]
122. Slater, B.; Smith, L. Outcrop Analogue for Trenton-Black River Hydrothermal Reservoirs, Mohawk Valley, New York. *AAPG Bull.* **2012**, *96*, 1369. [CrossRef]
123. Beauheim, R.L.; Roberts, R.M.; Avis, J.D. Hydraulic testing of low-permeability Silurian and Ordovician strata, Michigan Basin, southwestern Ontario. *J. Hydrol.* **2014**, *509*, 163–178. [CrossRef]
124. Sutcliffe, C.N.; Thibodeau, A.M.; Davis, D.W.; Al-Aasm, I.S.; Parmenter, A.; Zajacz, Z.; Jensen, M. Hydrochronology of a proposed deep geological repository for Low and Intermediate nuclear waste in southern Ontario from U-Pb dating of secondary minerals: Response to Alleghanian events. *Can. J. Earth Sci.* **2019**, *57*, 494–505. [CrossRef]

 *water*

Article

Origin of Drusy Dolomite Cement in Permo-Triassic Dolostones, Northern United Arab Emirates

Howri Mansurbeg [1,2,*], Mohammad Alsuwaidi [3], Shijun Dong [3], Salahadin Shahrokhi [4] and Sadoon Morad [3]

[1] General Directorate of Scientific Research Center, Salahaddin University-Erbil, Erbil 44001, The Kurdistan Region, Iraq
[2] School of the Environment, University of Windsor, Windsor, ON N9B 3P4, Canada
[3] Department of Earth Sciences, Khalifa University, Abu Dhabi 51133, United Arab Emirates; mohammad.alsuwaidi@ku.ac.ae (M.A.); dongshijun@yahoo.com (S.D.); sadoon.morad@ku.ac.ae (S.M.)
[4] URGA, s.r.o., Hydrogeologie, Geologie a Životní Prostředí, Sanace, 77200 Olomouc, Czech Republic; salah_sh1994@yahoo.com
* Correspondence: howri.mansurbeg@uwindsor.ca; Tel.: +45-42754220

Citation: Mansurbeg, H.; Alsuwaidi, M.; Dong, S.; Shahrokhi, S.; Morad, S. Origin of Drusy Dolomite Cement in Permo-Triassic Dolostones, Northern United Arab Emirates. *Water* **2021**, *13*, 1908. https://doi.org/10.3390/w13141908

Academic Editor: Kristine Walraevens

Received: 4 June 2021
Accepted: 6 July 2021
Published: 9 July 2021

Abstract: While the characteristics and origin of drusy calcite cement in carbonate deposits is well constrained in the literature, little attention is paid to drusy dolomite cement. Petrographic observations, stable isotopes, and fluid-inclusion microthermometry suggest that drusy dolomite cement in Permo-Triassic conglomerate/breccia dolostone beds in northern United Arab Emirates has precipitated as cement and not by dolomitization of drusy calcite cement. The low $\delta^{18}O_{VPDB}$ (−9.4‰ to −6.2‰) and high homogenization temperatures of fluid inclusions in drusy dolomite (T_h = 73–233 °C) suggest that dolomitization was caused by hot basinal brines (salinity = 23.4 wt% NaCl eq.). The $\delta^{13}C_{VPDB}$ values (+0.18‰ to +1.6‰) and $^{87}Sr/^{86}Sr$ ratio (0.708106 to 0.708147) indicate that carbon and strontium were derived from the host marine Permo-Triassic carbonates. Following this dolomitization event, blocky calcite (T_h = 148 °C; salinity = 20.8 wt% NaCl eq.) precipitated from the hot basinal brines. Unravelling the origin of drusy dolomite cement has important implications for accurate construction of paragenetic sequences in carbonate rocks and decipher the origin and chemistry of diagenetic waters in sedimentary basins.

Keywords: drusy dolomite; dolomitization; cementation; hot basinal fluids

1. Introduction

Calcite cement with a drusy mosaic texture, displaying crystal size growth from pore wall to pore center, is widely distributed in limestones [1,2]. Such cement commonly fills moldic pore and hence has been interpreted to be sourced by dissolution of aragonitic allochems [3–7]. The stable isotopic composition of drusy mosaic calcite cement coupled with the presence of moldic porosity have been interpreted to indicate a meteoric origin [8–10]. Dolomite cement in carbonate rocks commonly develops rhombic crystals that do not show systematic size variation within the pores [11]. Nevertheless, rare drusy mosaic dolomite cement has been reported in carbonate deposits too [12–15]. The origin of such dolomite is not fully explored, particularly regarding whether it is a primary precipitate or formed by dolomitization of precursor drusy calcite cement. The fluid flow and related diagenetic evolution, particularly dolomitization of carbonate succession in northern United Arab Emirates (UAE), is interpreted to be controlled by the tectonic evolution of the region [16,17].

In this paper, we constrain the origin of drusy mosaic dolomite cement in conglomerates/breccia of the Bih Formation (Permo-Triassic), Ras Al Khaima, northern UAE (Figure 1), using petrographic, stable isotopic, and fluid-inclusion microthermometric analyses. This study provides important insights into the conditions of formation of drusy dolomite cement and its associated diagenetic fluids. The presented data and discussions help in constructing more accurate paragenetic sequences in Permo-Triassic dolostones, northern UAE, and in other carbonate successions with similar depositional and diagenetic settings.

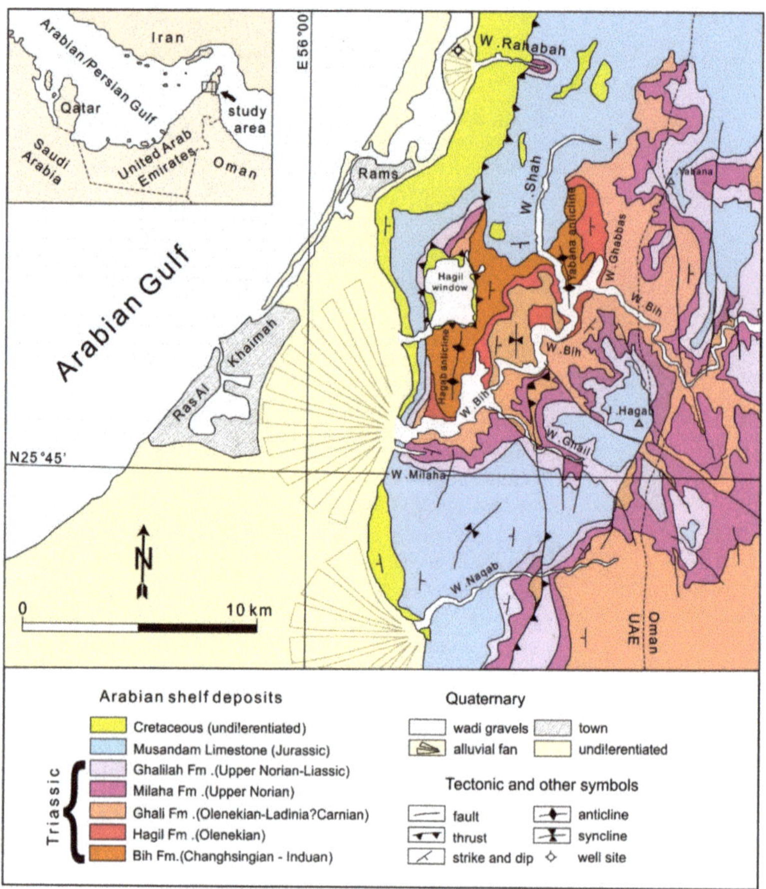

Figure 1. Location of the study area, northwest of the UAE (marked in green). Modified from [18–20].

2. Geological Setting

The Bih Formation (Permo-Triassic) crops out in the Musandam Peninsula, Ras Al Khaimah, northern United Arab Emirates (Figure 1). The formation is suggested to be age-equivalent to the Khuff Formation [19] which is an important hydrocarbon reservoir in the Arabian Gulf region. Deposition of the Bih Formation occurred in subtidal to intertidal (shoals, lagoon, and mudflats) environments on a passive margin of the Arabian plate [20].

The formation is divided into two units separated by a marker interval, which is called mid-Bih conglomerates/breccias [21]. The origin of this interval is uncertain, being interpreted to have resulted from the collapse of evaporite beds [20] tectonic deformation [20] or are lag deposits formed owing to marine transgression [22]. Strohmenger et al. [20] proposed that the Bih Formation corresponds to sequence KS4 through KS7 of the Khuff Formation (Figure 2). Two main tectonic events have affected the study area [23] including: (i) The obduction of Oman ophiolite on the Arabian platform during the Late Cretaceous. During this tectonic phase, the passive northeastern margin of the Tethys ocean became compressive and resulted in the close of the Tethys ocean [24]; (ii) the Zagros orogeny (Paleocene to Middle Miocene) [25–28] which caused large scale folding and thrusting [26].

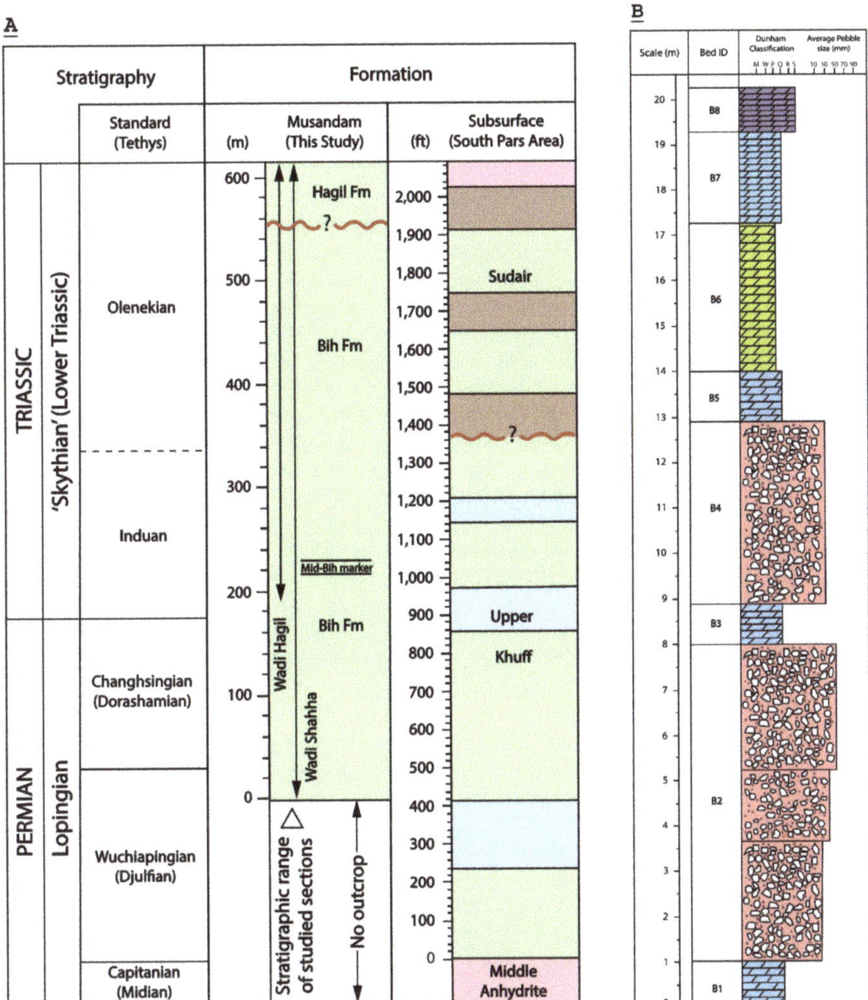

Figure 2. (A) Chronostratigraphic framework of the Khuff Formation in the subsurface compared to the stratigraphic range of the Bih Formation [21]. (B) Logging graph showing the thickness of conglomerate/breccia beds in the field.

3. Samples and Methods

Fifty-six samples were collected from the conglomerate/breccia beds and associated dolostone beds. Thin sections were prepared for all the samples after impregnation with blue epoxy. Representative polished uncovered thin sections were prepared and examined by a Technosyn Cold cathodoluminescence microscope (CL), backscattered electron imaging (BSEI), and energy dispersive X-ray analyses (EDS) attached to a Quanta scanning electron microscope. Fluid-inclusion microthermometry was conducted on two representative double polished 50 μm thick sections using a pre-calibrated LINKAM THMS600C stage fitted onto the Nikon E600 microscope.

Synthetic pure H_2O and CO_2 inclusions standard were used to calibrate the thermocouple. Both homogenization (T_h) and final ice melting temperatures (Tm_{ice}) were measured whenever the size of the inclusions permitted. The former is the minimum entrapment temperature of the inclusion (i.e., the temperature of mineral precipitation) [29]. The latter is the temperature at which the last ice melts in an aqueous fluid inclusion and reflects the trapped diagenetic fluid salinity [29]. The fluid salinity (wt% NaCl eq.) was inferred from Tm_{ice} using the Bodnar equation [30]. In order to avoid overheating, low heating rates were used for measuring homogenization temperatures (T_h).

Stable carbon and oxygen isotopes were obtained for the 36 micro-drilled drusy dolomite cement with a GVI IsoPrime continuous flow isotope ratio mass spectrometer system (CF-IRMS). The $\delta^{13}C$ and $\delta^{18}O$ isotopic values are determined on carbon dioxide (CO_2) from carbonate minerals by the reaction with 100% phosphoric acid (H_3PO_4) at 90 °C in a vacuum using standard procedures based on the principles reported in [31]. Samples and standards are weighed into Exetainer™ septum vials, sealed and flushed with pure helium. In addition, 100% phosphoric acid is injected into the vial, which is then placed in an aluminum tray maintained at 90 °C for more than 1 h. The CO_2 is extracted automatically with a double-hole needle Gilson auto-sampler connected to a GVI MultiFlow. The MultiFlow contains a 500-uL sample loop, and a GC column that separates CO_2, N_2, O_2, and H_2O. Helium carrier gas then transports the purified CO_2 into the GVI IsoPrime continuous flow isotope ratio mass spectrometer system (CF-IRMS) that measures the isotope ratios. Four representative micro-drilled drusy dolomite cement samples were analyzed for Sr isotopes. One mg of each sample is collected to react with suprapure HCl for 24 h. Purified Sr is obtained by chromatographic separation with 2.5 mL of AGW 50 × 8 (Biorad) cation exchange resin [32]. The analysis equipment is a Finnigan MAT 262 7-collector solid-source mass spectrometer with a single Re filament applying 1 μL of ionization enhancing solution [32].

4. Results

4.1. Field Observations

The base of the Bih Formation does not crop out, and the lowest stratigraphic part consists of a medium-to-thick-bedded brown dolomitized packstone–grainstones with locally well-developed coarse-crystalline dolostone and vuggy textures. About 180 m above the base of the exposed Bih Formation, the mid-Bih conglomerate/breccia occurs (approx. 20 m thick; Figures 2B and 3A). The conglomerate/breccia beds of Wadi Bih, which are interbedded with dolograinstones, are massive overall but locally show a low-angle, through cross stratification (Figure 3B). The beds are cross cut by a series of normal faults (Figure 3A). The basal contact between conglomerate/breccia and underlying dolograinstone beds is sharp and erosional (Figure 3C). The lower contact between dolograinstone beds with the conglomerate/breccia beds is marked by the presence of rip-up dolomudstone intraclasts (Figure 3D).

Stylolites occur within as well as along boundaries between the conglomerate/dolostone beds. Above the mid-Bih breccia, there are cycles of interbedded dolostones and dolomitic limestones, with algal laminations, bird-eye structures, and dissolution molds. The succession continues with 90 m cliff-forming, dark gray coarse-crystalline dolostones.

Figure 3. Field photographs showing: (**A**) The conglomerate/breccia beds are cut across by a series of normal faults, the conglomerate/breccia beds are shown by white lines, the faults are shown by orange lines. (**B**) Parallel lamination occurs at 14 cm above the basal contact in the conglomerate/breccia beds. (**C**) The contact (yellow line) between conglomerate/breccia and underlying massive dolostone bed. (**D**) The conglomerate/breccia beds have a gradational contact with overlying dolograinstone bed (arrows).

4.2. *Petrography of the Drusy Dolomite*

The conglomerate/breccia dolostone beds are cemented by drusy mosaic dolomite with a crystal size increasing progressively from 20 μm along the pore wall to 2 mm in the pore center (Figure 4A–D). In some cases, the centermost parts of the pore are occupied by blocky calcite (500–700 μm across). Dolomite crystals adjacent to this blocky calcite are partly calcitized (Figure 5A,B). The drusy dolomite crystals display sector CL zoning with alternating non-luminescent and bright orange luminescent zones (Figure 5C,D). The microcrystalline equant to bladed dolomite crystals (20–100 μm across) along the pore walls are dull luminescent. The blocky calcite crystals are dull luminescent with a local thin orange luminescent zone.

Dull luminescent calcite occurs along stylolites and fills fractures varying in width from 5 to 60 μm (Figure 5E,F). The stylolites (amplitudes up to 2 cm) and dissolution seams cut across the dolostone pebbles and have bed-parallel and bed-oblique orientations. Some of the stylolites are cutting across both pebbles and dolomite between pebbles (Figure 6A–D). The drusy mosaic dolomite has Fe and Mn contents below the EDS detection limits. Fe and Mn contents as low as few ppm can still generate zoning in dolomite [33–35].

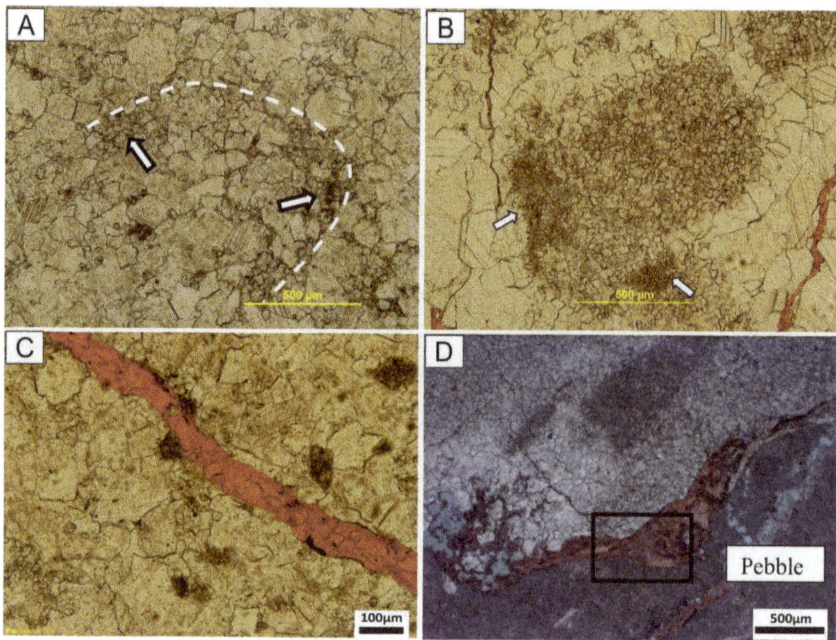

Figure 4. Optical photomicrographs (PPL) showing: (**A**) The fine dolomite crystals (arrows) in mosaic dolomite between pebbles are considered to be relics of fine-crystalline dolostone pebbles. The distribution of the fine dolomite crystals shows the rounded shape of pebble (dashed line). (**B**) The fine-crystalline dolostone pebble is partly replaced by the medium-crystalline dolomite (40 µm). Parts (arrows) of the pebble are not subjected to replacement. (**C**) Fracture cutting across the mosaic dolomite cement between pebbles was filled by the mosaic calcite cement (5 to 60 µm). The average width of this fracture is 70 µm. (**D**) Fracture-filling calcite along a pebble.

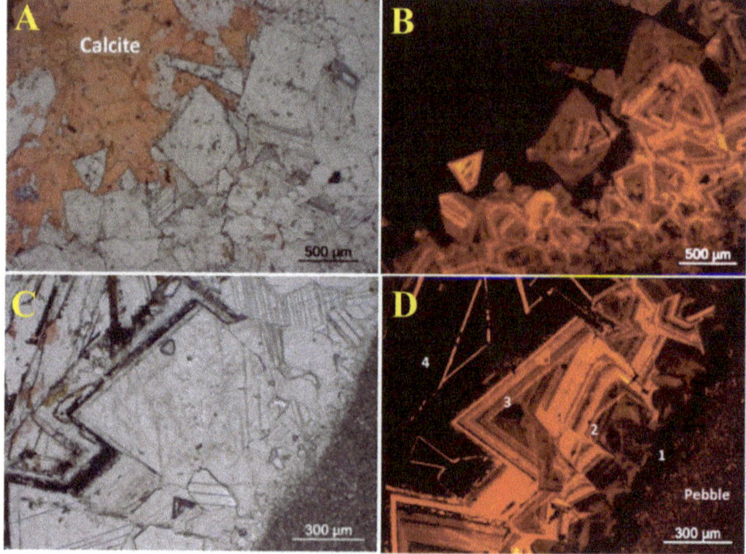

Figure 5. *Cont.*

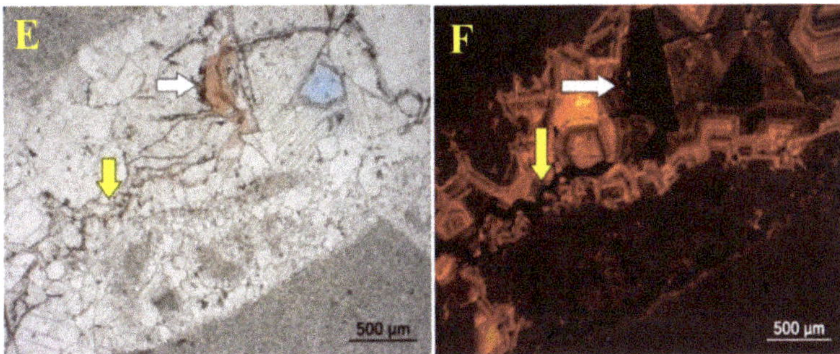

Figure 5. Optical (**A**; PPL) and companion CL (**B**) photomicrographs showing several generations of dolomite cement, which filled intergranular pores between pebbles are characterized by a distinguishable luminescent degree. Optical (**C**; PPL) and companion CL (**D**) photomicrographs showing: Coarse mosaic dolomite cement and coarse calcite cement, which is reddish stained in intergranular pores between pebbles; concentric growth pattern of planar-e to planar-s mosaic dolomite and non-luminescent calcite cements precipitated in pore space, where precursor dolomite was dissolved. Four generations of dolomite cement have been identified in the companion CL (**D**) photomicrograph. Optical (**E**; PPL) and companion CL photomicrographs (**F**) showing the calcite precipitated in intercrystalline pores between pebbles (white arrow) and within stylolite (yellow arrow) is characterized by a non-luminescent zone.

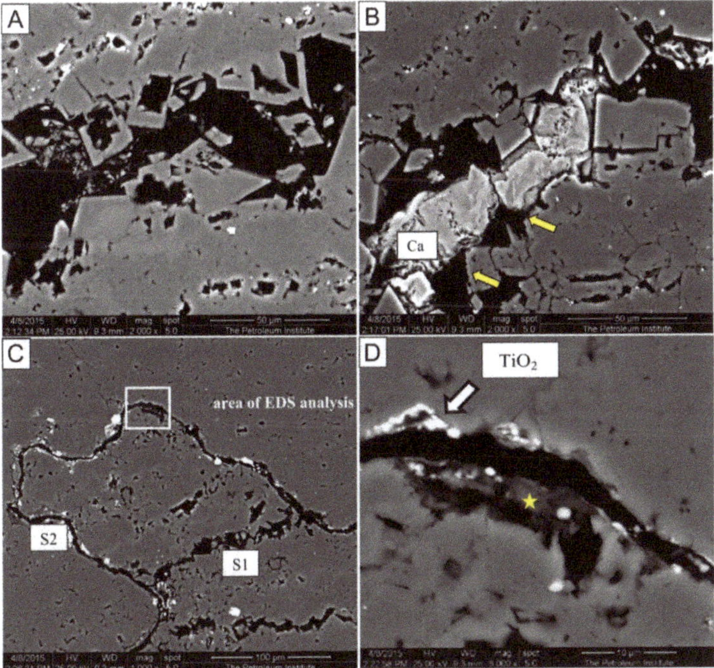

Figure 6. BSE images showing: (**A**) The presence of dolomite along a dissolution seam within a pebble. The core of rhombic dolomite crystals is dissolved while the rim is preserved. (**B**) Calcite cement (Ca) along the dissolution seam within the pebble, which precipitates after dissolution of dolomite (shown by an irregular boundary of dolomite crystal, yellow arrows). (**C**,**D**) stylolite/seams (S1 and S2) cutting across each other and the clay mineral and TiO_2 occur along the stylolite/solution seam within a pebble.

4.3. C, O, and Sr Isotopic Composition of Dolomite Cement

The $\delta^{13}C_{VPDB}$ values of drusy dolomite vary from +0.2‰ to +1.6‰ and $\delta^{18}O_{VPDB}$ from −9.4‰ to −6.2‰. The $\delta^{13}C_{VPDB}$ values of fracture-filling calcite cement vary between −3.8‰ and +1.9‰ and $\delta^{18}O_{VPDB}$ from −7.6‰ to −4.0‰. The $^{87}Sr/^{86}Sr$ ratios of two drusy and mosaic dolomite samples are 0.708106 and 0.708147, respectively (Figures 7 and 8A,B).

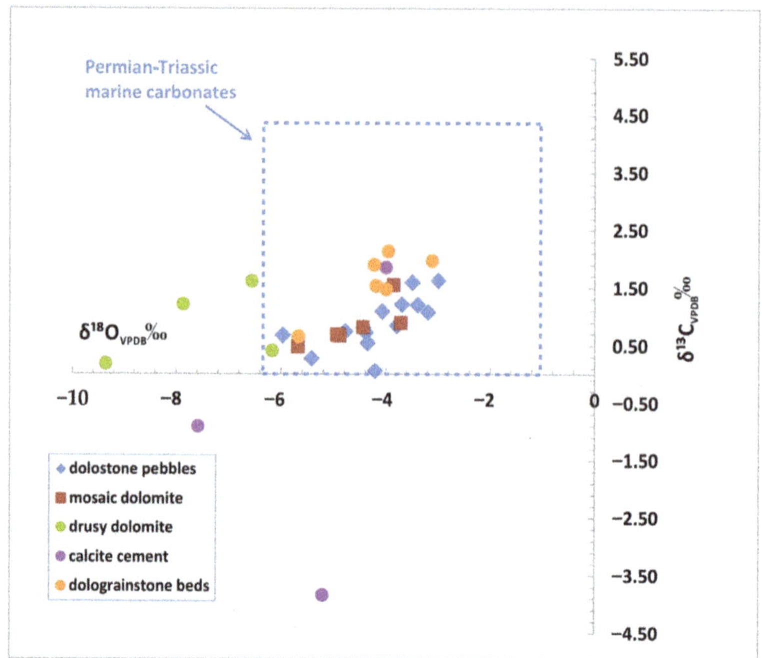

Figure 7. Cross plot of stable $\delta^{13}C_{VPDB}$ versus $\delta^{18}O_{VPDB}$ values of samples from the conglomerate/breccia beds and the dolograinstone beds in the conglomerate/breccia interval, as well as isotope values of Permian to Triassic marine carbonates. Stable oxygen isotope values of drusy dolomite are lower compared to other dolomites and slightly lower than sea coeval marine carbonates.

4.4. Fluid-Inclusion Microthermometry of Carbonate Cement

Fluid inclusions (1–20 µm across) in the drusy mosaic dolomite cement display two-phases (liquid and vapor). The crystal sizes of drusy dolomite are divided into fine (50–100 µm), medium (>100–500 µm), and coarse crystals (>500 µm). Results can be summarized as follows: (1) Homogenization temperatures (T_h) of fine-sized dolomite crystals (50–100 µm), range from 93 to 143 °C (av. 118 °C); (2) T_h of medium-sized dolomite crystals (100–500 µm) varies from 73 to 188 °C (av. 130.5 °C). The salinity of these fluid inclusions could not be measured owing to the small size of the inclusions (<2–3 µm); (3) T_h values of coarse crystalline dolomite (>500 µm) filling the centermost part of the pores vary from 118 to 233 °C (av. 175.5 °C), while salinity ranges from 21.8 to 25.1 wt% NaCl eq.; and (4) the calcite cement that postdates dolomite has a T_h value of 148 °C and salinity of 20.8 wt% NaCl eq. The distribution of T_h in calcite and drusy dolomite is shown in histograms (Figure 9).

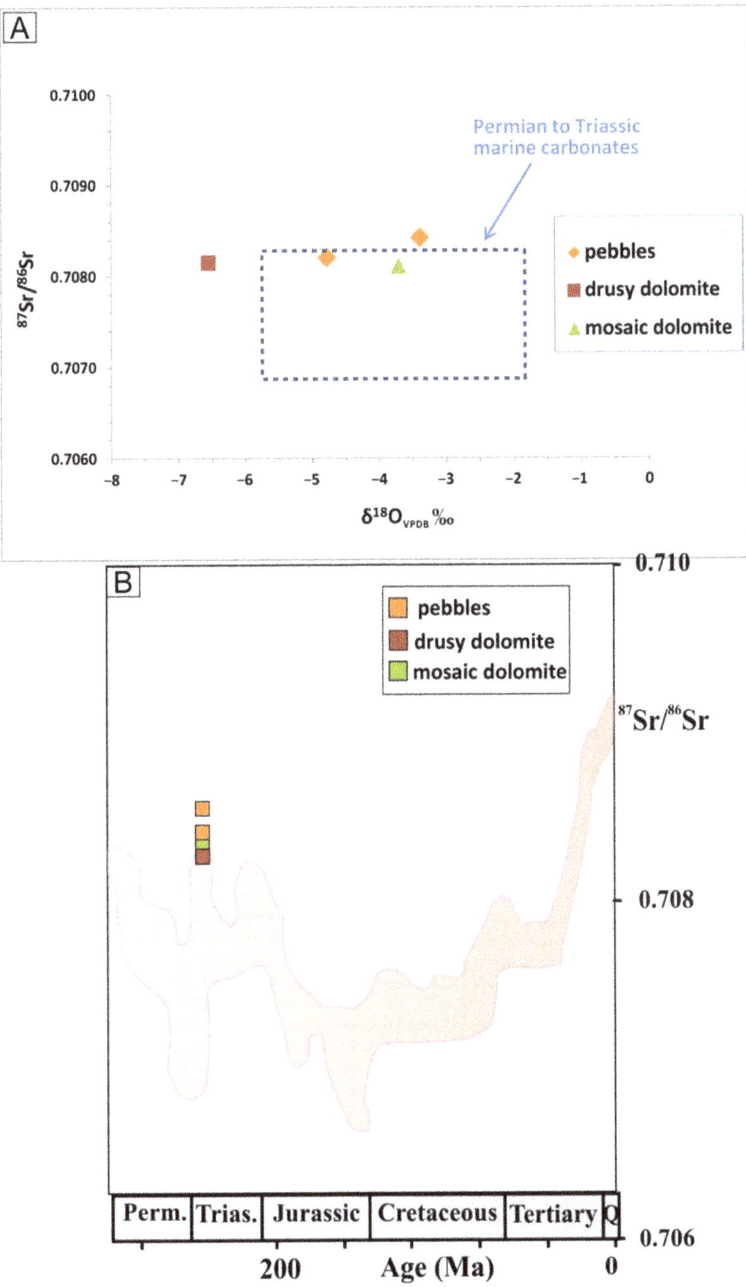

Figure 8. (**A**) Cross-plot of $^{87}Sr/^{86}Sr$ against $\delta^{18}O_{VPDB}$ showing two samples from the pebble have a slightly higher Sr isotope value than coeval marine carbonates. Two samples from drusy and mosaic dolomites are within the isotopic range of Permian to Triassic marine carbonates. The sample from drusy dolomite showing a lower $\delta^{18}O_{VPDB}$ value than coeval marine carbonates. (**B**) Measured $^{87}Sr/^{86}Sr$ ratios of different types of dolomite cements in the Bih Formation (Permo-Triassic) plotted on the sea water $^{87}Sr/^{86}Sr$ curve [31].

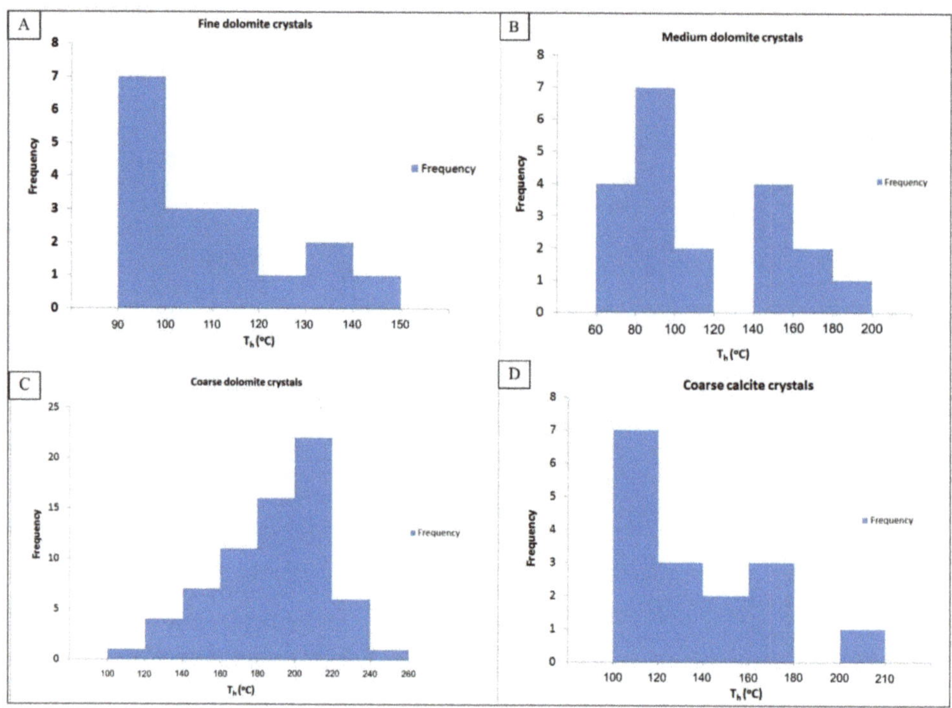

Figure 9. Histograms of fluid-inclusion homogenization temperatures (T_h) in fine-crystalline dolomite (**A**), medium-crystalline dolomite (**B**), coarse-crystalline dolomite (**C**), and calcite (**D**).

5. Discussion

Constraining the origin of the drusy dolomite in terms of whether it has been precipitated as cement, which is a common cement in limestones [13,14] or formed by the replacement of precursor drusy calcite [15] has implications for unravelling the role of diagenetic paleofluids, as well as for improvements in the definition of paragenetic sequences in carbonate rocks and related reservoir-quality features. There are two possible interpretations of the formation of drusy dolomite cement: (1) Precipitation as cement, and (2) fabric-preserving dolomitization of precursor drusy mosaic calcite. Evidence supporting precipitation as drusy dolomite includes the sector cathodoluminescence zoning and its close association with the latter precipitated blocky calcite.

Fluid-inclusion microthermometry of drusy mosaic dolomite and the subsequently precipitated calcite cement indicate formation from hot brines during compressional tectonic events in conjunction with the obduction of Oman ophiolites [16,36]. However, it is not immediately clear whether the drusy mosaic dolomite has been precipitated as cement [13] or formed by dolomitization of drusy mosaic calcite by the hot basinal brines [15,37]. Drusy calcite is a common cement in limestones [1,5–7,38]. Thus, the common presence of drusy mosaic dolomite in totally dolomitized limestones may suggest formation by fabric-preserving dolomitization [15]. The latter process suggests similar dissolution rates of drusy calcite and precipitation rate of dolomite during the dolomitization process. A similar fabric-preserving dolomite of Bih Formation has been reported in earlier studies of carbonate reservoirs in the world [39–41].

The high T_h (73 to 233 °C) and salinity (20.8 to 25.1 wt% NaCl eq.) of drusy mosaic dolomite and subsequently precipitated cement indicates that they have precipitated either during deep-burial diagenesis or by upward flux of hydrothermal/hot basinal fluids [42].

The increase in T_h with the increase in dolomite crystal size (av. 118 to 175.5 °C) is interpreted to indicate that the drusy dolomite precipitated with the increase in temperature during progressive burial. The maximum burial temperatures reached by the Bih Formation is approximately 190 to 200 °C [22]. The wide range of T_h of different carbonate phases indicates that some fluid inclusions have been subjected to re-equilibration during burial processes [29]. The relatively systematic T_h zoning from pore walls inwards (i.e., increase in crystal size) supports the formation of drusy dolomite by precipitation from pore fluids than by dolomitization of precursor drusy calcite cement. Apparently, the chemistry of the dolomitizing fluids has changed with time and reflected on the variation of Fe-rich and Fe-poor zoning of CL pattern of the drusy dolomite cement (cf., [2,36]).

The similar T_h (148 °C) and salinity values (20.8 wt% NaCl eq.) of late blocky calcite and fracture-filling medium to coarse calcite cement is attributed to the circulation of hot brines [17]. The considerably lower $\delta^{13}C_{VPDB}$ values (−3.8‰ and −0.9‰) of fracture-filling calcite cement than drusy dolomite samples might indicate the deviation of dissolved carbon from degradation of hydrocarbons or organic matter [42]. Using the $\delta^{18}O_{VPDB}$ values (−9.4‰ to −6.2‰) and T_h (73 to 233 °C) of drusy dolomite, the fractionation equation of Land [43] it is inferred that precipitation has occurred from evolved brines ($\delta^{18}O_{VSMOW}$ = +0.29‰ to +10.96‰). The wide extent of variations in the degree of geochemical evolution of the brines could be related to variable fluid sources and variable degrees of interaction with the host rocks in the basin. The positive correlation between C and O isotopes of drusy mosaic dolomite (Figure 7) suggests increasing input of ^{12}C into the formation waters with the increasing temperature [44]. The small variations in Sr isotopic compositions between the various types of carbonate cements suggest a related origin (Figure 8A,B). The $^{87}Sr/^{86}Sr$ ratios in one of the pebbles (0.708202) is within the isotopic range of Permian to Triassic seawater values (0.7069 to 0.7083) [31]. The slightly higher ratio in another pebble sample (0.708424) than the range for Permo-Triassic seawater (Figure 8B) suggests that the hot basinal fluids have interacted with siliciclastic rocks [45,46]. The similar carbon isotope and $^{87}Sr/^{86}Sr$ ratio of drusy dolomite and coeval marine carbonates ([31]; 0.708106 and 0.708147, respectively) suggests that dissolved carbon and Sr were derived from the dissolution of marine carbonates and/or marine pore waters. The slightly lower oxygen isotope value of drusy dolomite than coeval marine carbonates (Figure 8A) is attributed to precipitation at elevated temperatures.

The development of stylolites and dissolution seams along the rim of pebbles or cutting across both pebbles and dolomite between pebbles indicate that stylolitization postdates the drusy dolomite cement. This complex relationship between fracturing, stylolitization, and dolomite cementation may be caused by multiple phases of tectonic movements. The cross-cutting of low amplitude seams by high amplitude stylolites in conglomerate/breccia samples could be a result of increasing compressive stress by tectonic movements generated during the obduction of Oman ophiolites [47,48]. The presence of dolomite cement (T_h up to 190 °C; salinity up to 20.8 wt% NaCl eq.) along the stylolites suggests the flux of hot basinal brines and may indicate that the dolomite was precipitated by dolomitization of the host limestones prior to stylolitization. Dolomite cement apparently has been concentrated in the vicinity of the stylolitization surface as an insoluble residue (cf., [49]). Similar observations and interpretations have been presented for carbonate successions elsewhere [49–51]. The presence of clay minerals (locally pigmented by TiO_2) along stylolites suggests that these minerals promoted the pressure dissolution of carbonates [52,53].

Constraining the depths at which the flux of hot dolomitizing basinal brines occurred is difficult without radiometric dating of dolomite [54,55]. However, the cross-cutting relationship between the stylolites and dolomite cement suggests that dolomitization took place at shallow depths [36].

6. Conclusions

The puzzling origin of pore-filling drusy mosaic dolomite cement in Permo-Triassic conglomerate/breccia beds outcropping in northern United Arab Emirates is constrained using the integrated field, petrographic, isotopic, and fluid-inclusion microthermometric analyses. There are two possible interpretations of the formation of drusy dolomite cement: (1) Precipitation as cement, and (2) fabric-preserving dolomitization of precursor drusy mosaic calcite. Fluid-inclusion microthermometry indicates the formation of this dolomite from hot basinal brines (T_h = 73 to 233 °C; salinity = 21.8% to 25.1% wt% NaCl eq.). Using the average homogenization temperature and oxygen isotope of dolomite $\delta^{18}O_{VPDB}$ (−9.4‰ to −6.2‰) the brines are inferred to be geochemically evolved with $\delta^{18}O_{VSMOW}$ (+0.29‰ to +10.96‰). Lines of evidence supporting precipitation as drusy dolomite from pore waters include the sector cathodoluminescence zoning, the relatively systematic T_h zoning, and with the presence of the latter precipitated blocky calcite, which has also precipitated from hot basinal brines (T_h = 148 °C salinity = 20.8 wt% NaCl eq.). The construction of accurate paragenetic sequences is important for establishing exploration and production models for carbonate reservoirs. To achieve this goal, diagenesis should be linked to fluid history and tectonic evolution of the basin.

Author Contributions: H.M.: Conceptualization, Writing—original draft. M.A.: Writing—review & editing. S.D.: Formal analysis, Methodology. S.S.: Writing—review & editing, Visualization. S.M.: Writing—review & editing, Supervision, Resources, Investigation. All authors have read and agreed to the published version of the manuscript.

Funding: This research received no external funding.

Institutional Review Board Statement: Not applicable.

Informed Consent Statement: Not applicable.

Data Availability Statement: Not applicable.

Acknowledgments: The fluid inclusion analyses were performed by Daniel Morad at the French Petroleum Institute, Paris, France, the carbon and oxygen isotope analyses at the Environmental Isotope laboratory, University of Waterloo, Canada, and the strontium isotope analyses at Bochum University, Germany.

Conflicts of Interest: The authors declare no conflict of interest.

References

1. Bathurst, R.G.C. Diagenesis in Mississippian Calcilutites and Pseudobreccias. *J. Sediment. Res.* **1959**, *29*, 365–376. [CrossRef]
2. Tucker, M.E.; Wright, V.P. Diagenetic Processes, Products and Environments. *Carbonate Sedimentol.* **2009**, *482*, 314–364. [CrossRef]
3. Friedman, G.M. Early Diagenesis and Lithification in Carbonate Sediments. *J. Sediment. Res.* **1964**, *34*, 777–813. [CrossRef]
4. Gavish, E.; Friedman, G.M. Progressive diagenesis in Quaternary to Late Tertiary carbonate sediments: Sequence and time scale. *J. Sediment. Res.* **1969**, *39*, 980–1006.
5. Dodd, J.R. Processes of Conversion of Aragonite to Calcite with Examples from the Cretaceous of Texas. *J. Sediment. Res.* **1966**, *36*, 36. [CrossRef]
6. Rousseau, M.; Dromart, G.; Garcia, J.-P.; Atrops, F.; Guillocheau, F. Jurassic evolution of the Arabian carbonate platform edge in the central Oman Mountains. *J. Geol. Soc.* **2005**, *162*, 349–362. [CrossRef]
7. Tackett, L.S.; Bottjer, D.J. Paleoecological succession of norian (late triassic) benthic fauna in eastern panthalassa (luning and gabbs formations, west-central nevada). *PALAIOS* **2016**, *31*, 190–202. [CrossRef]
8. Lohmann, K.C.; Walker, J.C.G. Secular variation in 13C and 16O compositTion of phamerozoic oceans. *Geol. Soc. Am. Abstr. Prog.* **1984**, *16*, 578.
9. Carpenter, S.J.; Lohmann, K.C. Delta 18 and delta 13 C variations in Late Devonian marine cements from the Golden Spike and Nevis reefs, Alberta, Canada. *J. Sediment. Res.* **1989**, *59*, 792–814. [CrossRef]
10. Scholle, P.A.; Ulmer-Scholle, D.S. *A Color Guide to the Petrography of Carbonate Rocks: Grains, Textures, Porosity, Dia-Genesis, AAPG Memoir 77*; AAPG: Tulsa, OK, USA, 2003.
11. Morad, S.; Al-Aasm, I.S.; Nader, F.H.; Ceriani, A.; Gasparrini, M.; Mansurbeg, H. Impact of diagenesis on the spatial and temporal distribution of reservoir quality in the Jurassic Arab D and C members, offshore Abu Dhabi oilfield, United Arab Emirates. *GeoArabia* **2010**, *17*, 17–56.
12. Mattavelli, L.; Chilingarian, G.; Storer, D. Petrography and diagenesis of the taormina formation, gela oil field, sicily (Italy). *Sediment. Geol.* **1969**, *3*, 59–86. [CrossRef]
13. Chatalov, A.; Bonev, N.; Ivanova, D. Depositional characteristics and constraints on the mid-Valanginian demise of a car-bonate platform in the intra-Tethyan domain, Circum- Rhodope Belt, northern Greece. *Cretaceous Res.* **2015**, *55*, 84–115. [CrossRef]

14. Gariboldi, K.; Gioncada, A.; Bosio, G.; Malinverno, E.; Di Celma, C.; Tinelli, C.; Cantalamessa, G.; Landini, W.; Urbina, M.; Bianucci, G. The dolomite nodules enclosing fossil marine vertebrates in the East Pisco Basin, Peru: Field and petrographic insights into the Lagerstätte formation. *Palaeogeogr. Palaeoclim. Palaeoecol.* **2015**, *438*, 81–95. [CrossRef]
15. Mansurbeg, H.; Morad, D.; Othman, R.; Morad, S.; Ceriani, A.; Al-Aasm, I.; Kolo, K.; Spirov, P.; Proust, J.N.; Preat, A.; et al. Hydrothermal dolomitization of the Bekhme formation (Upper Cretaceous), Zagros Basin, Kurdistan Region of Iraq: Record of oil migration and degradation. *Sediment. Geol.* **2016**, *341*, 147–162. [CrossRef]
16. Callot, J.P.; Breesch, L.; Guilhaumou, N.; Roure, F.; Swennen, R.; Vilasi, N. Paleo-fluids characterisation and fluid flow modelling along a regional transect in Northern United Arab Emirates (UAE). *Arab. J. Geosci.* **2010**, *4*, 413–437. [CrossRef]
17. Fontana, S.; Nader, F.; Morad, S.; Ceriani, A.; Daniel, J.-M.; Mengus, J.-M.; Al-Aasm, I.S. Fluid-rock interactions associated with regional tectonics and basin evolution. *Sedimentology* **2013**, *61*, 660–690. [CrossRef]
18. Maurer, F.; Rettori, R.; Martini, R. Triassic stratigraphy, facies and evolution of the Arabian shelf in the northern United Arab Emirates. *Int. J. Earth Sci.* **2008**, *97*, 765. [CrossRef]
19. Searle, M.P. Thrust tectonics of the Dibba zone and the structural evolution of the Arabian continental margin along the Musandam Mountains (Oman and United Arab Emirates). *J. Geol. Soc.* **1988**, *145*, 43–53. [CrossRef]
20. Strohmenger, C.J.; Alway, R.H.; Broomhall, R.W.; Hulstrand, R.F.; Al-Mansoori, A.; Abdalla, A.A.; Al-Aidarous, A. Sequence stratigraphy of the Khuff Formation comparing subsurface and outcrop data. In Proceedings of the Abu Dhabi International Petroleum Exhibition and Conference, Abu Dhabi, United Arab Emirates, 13 October 2002.
21. Maurer, F.; Martini, R.; Rettori, R.; Hillgärtner, H.; Cirilli, S. The geology of Khuff outcrop analogues in the Musandam Peninsula, United Arab Emirates and Oman. *GeoArabia* **2009**, *14*, 125–158.
22. Fontana, S.; Morad, S.; Nader, F.H.; Ceriani, A.; Al-Aasm, I.S. Breccia beds in the Khuff (Permian–Triassic) in Ras Al Khaimah, United Arab Emirates: Collapse or transgressive in origin? In the Permo–Triassic Sequence of the Arabian Plate, Abstracts of the EAGE's Third Arabian Plate Geology Workshop, Kuwait. Abstract. *GeoArabia J. Middle East Pet. Geosci-Ences.* **2012**, *17*, 225–230.
23. Fontana, S.; Morad, S.; Nader, F.H.; Al-Aasm, I.S.; Ceriani, A. Diagenesis of the Permo-triassic Succession of the Northeastern UAE, an Outcrop Analogue of the Khuff Formation. In Proceedings of the 73rd EAGE Conference and Exhibition Incorporating SPE EUROPEC, Vienna, Austria, 23–26 May 2011; p. 238. [CrossRef]
24. Warburton, J.; Burnhill, T.; Graham, R.; Isaac, K. *The Evolution of the Oman Mountains Foreland Basin*; Geological Society, Special Publications: London, UK, 1990; Volume 49, pp. 419–427.
25. Breesch, L.; RSwennen, B.; Dewever, F.; Roure, B. Vincent. Diagenesis and fluid system evolution in the northern Oman Mountains, United Arab Emirates: Implications for petroleum exploration. *GeoArabia* **2011**, *16*, 111–148.
26. Breesch, L.; Swennen, R.; Vincent, B. Fluid flow reconstruction in hanging and footwall carbonates: Compartmentalization by Cenozoic reverse faulting in the Northern Oman Mountains (UAE). *Mar. Pet. Geol.* **2009**, *26*, 113–128. [CrossRef]
27. Michaelis, P.; Pauken, R. *Seismic Interpretation of the Structure and Stratigraphy of the Strait of Hormuz*; Geological Society, Special Publications: London, UK, 1990; Volume 49, pp. 387–395.
28. Searle, M.P.; James, N.P.; Calon, T.J.; Smewing, J.D. Sedimentological and structural evolution of the Arabian continental margin in the Musandam Mountains and Dibba zone, United Arab Emirates. *GSA Bull.* **1983**, *94*, 1381. [CrossRef]
29. Goldstein, R.H.; Reynolds, T.J. Systematics of fluid inclusions in diagenetic minerals. *SEPM Short Course* **1994**, *31*, 199.
30. Bodnar, R. Revised equation and table for determining the freezing point depression of H2O-Nacl solutions. *Geochim. Cosmochim. Acta* **1993**, *57*, 683–684. [CrossRef]
31. Veizer, J.; Ala, D.; Azmy, K.; Bruckschen, P.; Buhl, D.; Bruhn, F.; Godderis, Y. 87Sr/86Sr, δ13C and δ18O evolution of Phan-erozoic seawater. *Chem. Geol.* **1999**, *161*, 59–88. [CrossRef]
32. Epstein, S.; Buchsbaum, R.; Lowenstam, H.A.; Urey, H.C. Revised carbonate-water Isotopic Temperature Scale. *Geol. Soc. Am. Bull.* **1953**, *64*, 1315–1326. [CrossRef]
33. Savard, M.M.; Veizer, J.; Hinton, R. Cathodoluminescene at low Fe and Mn concentrations; a SIMS study of zones in natural calcites. *J. Sediment. Res.* **1995**, *65*, 208–213. [CrossRef]
34. Pierson, B.J. The control of cathodoluminescence in dolomite by iron and manganese. *Sedimentology* **1981**, *28*, 601–610. [CrossRef]
35. Carmichael, S.K.; Ferry, J.M. Formation of replacement dolomite in the Latemar carbonate buildup, Dolomites, northern Italy: Part Origin of the dolomitizing fluid and the amount and duration of fluid flow. *Am. J. Sci.* **2008**, *308*, 885–904. [CrossRef]
36. Fontana, S.; Nader, F.H.; Morad, S.; Ceriani, A.; Al-Aasm, I.S. Diagenesis of the Khuff Formation (Permian–Triassic), Northern United Arab Emirates. In *Frontiers in Earth Sciences*; Springer: Berlin/Heidelberg, Germany, 2013; pp. 203–220.
37. Philip, J.M.; Gari, J. Late Cretaceous heterozoan carbonates: Palaeoenvironmental setting, relationships with rudist carbonates (Provence, south-east France). *Sediment. Geol.* **2005**, *175*, 315–337. [CrossRef]
38. Fairchild, I.J.; Humbrey, M.J. The Vendian succession of northeastern Spitsbergen: Petrogenesis of a dolomite-tillite association. *Precambrian Res.* **1984**, *26*, 111–167. [CrossRef]
39. Kimbell, T.N. Sedimentology and Diagenesis of Late Pleistocene Fore-Reef Calcarenites, Barbados, West Indies: A Geo-Chemical and Petrographic Investigation of Mixing Zone Diagenesis. Ph.D. Thesis, The University of Texas at Dallas, Dallas, TX, USA, 1993; p. 296.
40. Cantrell, D.; Peter, S.; Royal, H. Genesis and characterization of dolomite, Arab-D reservoir, Ghawar field, Saudi Arabia. *GeoArabia* **2004**, *9*, 11–36.

41. Hutsky, A.J.; Fielding, C.R.; Frank, T.D. Evidence for a petroleum subsystem in the Frontier Formation of the Uinta–Piceance Basin petroleum province. *AAPG Bull.* **2016**, *100*, 1033–1059. [CrossRef]
42. Morad, S.; De Ros, L.F.; Nystuen, J.P.; Bergan, M. Carbonate Diagenesis and Porosity Evolution in Sheet-Flood Sandstones: Evidence from the Middle and Lower Lunde Members (Triassic) in the Snorre Field, Norwegian North Sea. In *Carbonate Cementation in Sandstones*; Wiley: Hoboken, NJ, USA, 2009; Volume 26, pp. 53–85.
43. Land, L.S. The application of stable isotopes to studies of the origin of dolomite and to problems of diagenesis of clastic sedi-ments. In *Stable Isotopes in Sedimentary Geology*; Arthur, M.A., Anderson, T.F., Kaplan, I.R., Veizer, J., Land, L.S., Eds.; Special Publications of SEPM: Tulsa, OK, USA, 1983; Volume 10, pp. 41–422.
44. Morad, S. *Carbonate Cementation in Sandstones: Distribution Patterns and Geochemical Evolution (Special Publication 26 of the IAS)*; John Wiley & Sons: Hoboken, NJ, USA, 2008; p. 72.
45. Davies, G.R.; Smith, L.B. Structurally controlled hydrothermal dolomite reservoir facies: An overview. *AAPG Bull.* **2006**, *90*, 1641–1690. [CrossRef]
46. Gasparrini, M.; Bechstädt, T.; Boni, M. Massive hydrothermal dolomites in the southwestern Cantabrian Zone (Spain) and their relation to the Late Variscan evolution. *Mar. Pet. Geol.* **2006**, *23*, 543–568. [CrossRef]
47. Sirat, M.; Al-Aasm, I.; Morad, S.; Aldahan, A.; Al-Jallad, O.; Ceriani, A.; Morad, D.; Mansurbeg, H.; Al Suwaidi, A. Saddle dolomite and calcite cements as records of fluid flow during basin evolution: Paleogene carbonates, United Arab Emirates. *Mar. Pet. Geol.* **2016**, *74*, 71–91. [CrossRef]
48. Morad, D.; Nader, F.; Gasparrini, M.; Morad, S.; Rossi, C.; Marchionda, E.; Al Darmaki, F.; Martines, M.; Hellevang, H. Comparison of the diagenetic and reservoir quality evolution between the anticline crest and flank of an Upper Jurassic carbonate gas reservoir, Abu Dhabi, United Arab Emirates. *Sediment. Geol.* **2018**, *367*, 96–113. [CrossRef]
49. Paganoni, M.; Al Harthi, A.; Morad, D.; Morad, S.; Ceriani, A.; Mansurbeg, H.; Al Suwaidi, A.; Al-Aasm, I.S.; Ehrenberg, S.N.; Sirat, M. Impact of stylolitization on diagenesis of a Lower Cretaceous carbonate reservoir from a giant oilfield, Abu Dhabi, United Arab Emirates. *Sediment. Geol.* **2016**, *335*, 70–92. [CrossRef]
50. Braithwaite, C. Stylolites as open fluid conduits. *Mar. Pet. Geol.* **1989**, *6*, 93–96. [CrossRef]
51. Neilson, J.E.; Oxtoby, N.H.; Simmons, M.D.; Simpson, I.R.; Fortunatova, N.K. The relationship between petroleum emplace-ment and carbonate reservoir quality: Examples from Abu Dhabi and the Amu Darya Basin. *Mar. Pet. Geol.* **1998**, *15*, 57–72. [CrossRef]
52. Aharonov, E.; Katsman, R. Interaction between pressure solution and clays in stylolite development: Insights from modeling. *Am. J. Sci.* **2009**, *309*, 607–632. [CrossRef]
53. Ehrenberg, S.N.; Morad, S.; Yaxin, L.; Chen, R. Stylolites and Porosity in A Lower Cretaceous Limestone Reservoir, Onshore Abu Dhabi, U.A.E. *J. Sediment. Res.* **2016**, *86*, 1228–1247. [CrossRef]
54. Salih, N.; Mansurbeg, H.; Kolo, K.; Préat, A. Hydrothermal Carbonate Mineralization, Calcretization, and Microbial Diagenesis Associated with Multiple Sedimentary Phases in the Upper Cretaceous Bekhme Formation, Kurdistan Region-Iraq. *Geoscience* **2019**, *9*, 459. [CrossRef]
55. Mansurbeg, H.; Alsuwaidi, M.; Salih, N.; Shahrokhi, S.; Morad, S. Integration of stable isotopes, radiometric dating and mi-crothermometry of saddle dolomite and host dolostones (Cretaceous carbonates, Kurdistan, Iraq): New insights into hydro-thermal dolomitization. *Mar. Pet. Geol.* **2021**, *127*, 104989. [CrossRef]

 *water*

Article

The Origin of Quartz Cement in the Upper Triassic Second Member of the Xujiahe Formation Sandstones, Western Sichuan Basin, China

Jie Ren [1,2], Zhengxiang Lv [1,*], Honghui Wang [1,3], Jianmeng Wu [2] and Shunli Zhang [4]

1 College of Energy Resources, Chengdu University of Technology, Chengdu 610059, China; rj2018cdut@163.com (J.R.); wanghh202013@163.com (H.W.)
2 Sinopec Matrix Xinan MWD/LWD & Logging Corporation, Chengdu 610100, China; wujm.osxn@sinopec.com
3 Sichuan University of Science & Engineering, Zigong 643000, China
4 Exploration and Development Research Institute, Sinopec Southwest Company, Chengdu 610041, China; zhangshunli91@163.com
* Correspondence: lvzhengxiang13@cdut.cn

Abstract: High-precision in situ $\delta^{18}O$ values obtained using secondary ion mass spectrometry (SIMS) for µm-size quartz cement are applied to constrain the origin of the silica in the deep-buried Upper Triassic second member of Xujiahe Formation tight sandstones, western Sichuan Basin, China. Petrographic, cathodoluminescence (CL), and fluid inclusion data from the quartz cements in the Xu2 sandstones indicate three distinct, separate quartz precipitation phases (referred to as Q1, Q2, and Q3). The Q1 quartz cement was formed at temperatures of approximately 56–85 °C and attained the highest $\delta^{18}O$ values (ranging from 18.3 to 19.05‰ Vienna Standard Mean Ocean Water (VSMOW)). The Q2 quartz cement was generated at temperatures of approximately 90–125 °C, accompanying the main phase of hydrocarbon fluid inclusions, with the highest Al_2O_3 content and high $\delta^{18}O$ values (ranging from 15 to 17.99‰ VSMOW). The Q3 quartz cement was formed at temperatures of approximately 130–175 °C, with the lowest $\delta^{18}O$ values (ranging from 12.79 to 15.47‰ VSMOW). A portion of the Q2 and Q3 quartz cement has a relatively high K_2O content. The dissolution of feldspar and volcanic rock fragments was likely the most important source of silica for the Q1 quartz cement. The variations in $\delta^{18}O_{(water)}$ and trace element composition from the Q2 quartz cement to the Q3 quartz cement suggest that hydrocarbon emplacement and water-rock interactions greatly altered the chemistry of the pore fluid. Feldspar dissolution by organic acids, clay mineral reactions (illitization and chloritization of smectite), and pressure dissolution were the main sources of silica for the Q2 and Q3 quartz cements, while transformation of the clay minerals in the external shale unit was a limited silica source.

Keywords: quartz cement; oxygen isotopes; silica sources; tight sandstones; Sichuan Basin

Citation: Ren, J.; Lv, Z.; Wang, H.; Wu, J.; Zhang, S. The Origin of Quartz Cement in the Upper Triassic Second Member of the Xujiahe Formation Sandstones, Western Sichuan Basin, China. *Water* **2021**, *13*, 1890. https://doi.org/10.3390/w13141890

Academic Editor: Howri Mansurbeg

Received: 14 May 2021
Accepted: 3 July 2021
Published: 8 July 2021

1. Introduction

Quartz cement is the volumetrically most significant diagenetic mineral altering porosity in deep-buried sandstones, which occurs in mature, medium- to coarse-grained quartz arenites, typical of tectonically stable cratonic basins [1–3]. Understanding the origin and distribution of quartz cement is consequently of economic importance for reservoir quality prediction in oilfield sandstones [4]. Although authigenic quartz cements have been extensively studied, the pH, temperature, pressure, detrital composition, oil and gas emplacement, and chlorite coating play an important role in the precipitation of authigenic quartz, but the formation mechanism and accurate formation time of authigenic quartz are not fully understood [3,5–7]. A certain volume of silica is potentially liberated from pressure solution at grain contacts and through feldspar dissolution, the transformation of smectite into illite or the recrystallization of biogenic silica. Silica solutions are also introduced into sand bodies from external sources through faults and fractures or by advection from deep underlying reservoirs or from adjacent shale units [3,7–10]. The quartz cements in most sedimentary basins were formed at diagenetic temperatures ranging from 60–145 °C, at burial depths greater than 2 km, and some scholars have hypothesized that temperature-controlled precipitation kinetics are the main control on quartz cement formation and that quartz precipitation exponentially increases with the temperature once the kinetic barriers are overcome above the cement threshold (70–80 °C) [2,5,11–13].

Quartz cement is most abundant in the deep-buried Upper Triassic second member of the Xujiahe Formation tight gas sandstones. Many studies have thus attempted to evaluate the temperature and timing of precipitation as well as the silica origin, typically through the combined application of quartz fluid inclusion and isotopic analysis methods [9,10,14,15]. When the temperature of authigenic quartz precipitation is known, the oxygen isotope composition of the fluid from which silica was precipitated can be calculated from the measured oxygen isotope ratios of the quartz cement [16]. In this study, we perform high-precision, high-spatial-resolution secondary ion mass spectrometry (SIMS) analysis of the quartz cement in the second member of the Xujiahe Formation sandstones. With an ion microprobe, individual quartz overgrowths and pore filling quartz cement are analyzed in situ at a spatial resolution of 20–30 μm. The objective of this study is to interpret the origin and transport mechanism of silica in addition to the nature and composition of the fluids that precipitated the quartz cement in the second member of the Xujiahe Formation sandstones.

2. Geologic Setting

The Sichuan Basin is a large oil-bearing superimposed basin. The study area is located in front of the Longmenshan Orogenic Belt, which represents the boundary between the western edge of the Yangtze plate and the Songpan-Ganzi fold (Figure 1a,b). The western Sichuan Basin, in which the study area is located (Figure 1c), is tectonically subdivided into the Zitong depression, Xiaoquan-Fenggu structural zone, Dayi-Yazihe-Anxian fault folding zone, Chengdu depression, and Zhixinchang-Luodai structural zone [17–19].

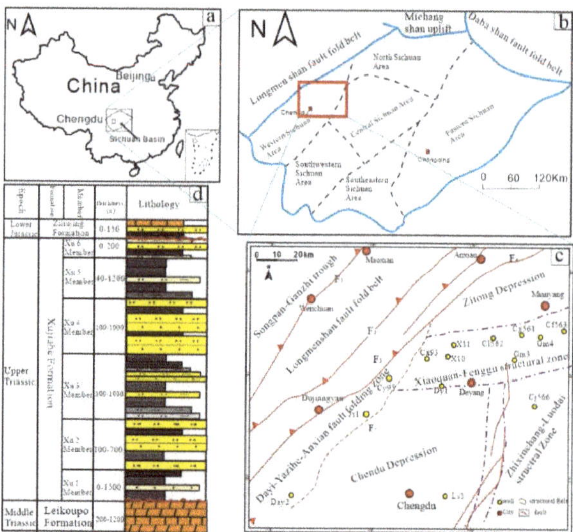

Figure 1. (**a**) The location of the Sichuan Basin; (**b**) Geological setting of the Sichuan Basin; (**c**) The tectonic divisions of the western Sichuan Basin; (**d**) Generalized stratigraphy and lithological types of the western Sichuan Basin.

The study area mainly experienced the following two tectonic evolution stages: the Sinian–Middle Triassic tectonic evolution of a passive continental margin and the Late Triassic–Eocene evolution of a foreland basin [20,21]. The Indosinian movement at the end of the Late Triassic resulted in the collision of the Yangtze plate with the North China plate. The Longmenshan island chain along the western edge of the Sichuan Basin gradually evolved into a strong thrust nappe orogenic belt and entered the stage of foreland basin evolution, and the coal-bearing clastic rocks of the Xujiahe Formation were deposited [17,18,22–24].

The Xujiahe Formation in the Sichuan Basin unconformably overlies the Middle Triassic Leikoupo Formation, which consists of dolomite, dolomitic limestone, and limestone and is also conformably overlain by the Lower Jurassic Ziliujing Formation. From bottom to top, the Xujiahe Formation in the western Sichuan Basin can be divided into five members, namely, Xu1–Xu5 (Figure 1d). The Xu1, Xu3 and Xu5 members mainly consist of black shale, mudstone, and coal seams and are regional petroleum source rocks for the Xujiahe Formation, whereas the Xu2 and Xu4 members are predominantly sandstones, which are an important sandstone reservoir and production layer, respectively, in the western Sichuan Basin [19,25]. The burial depth of the Xu2 member is more than 4600 m, and the sedimentary thickness is between 560 and 660 m. The Xu2 member was largely deposited in a marine delta front belonging to a marine-to-continental transitional environment, mainly consisting of distributary channel and mouth bar sandstones [19,23,26].

3. Materials and Methods

A total of 102 samples were collected from 8 wells in the Xu2 member in the western Sichuan Basin, at burial depths ranging from 4800–5200 m. Eighty-six samples were selected for thin section observation, and thin sections were prepared for multiple purposes via blue-dyed epoxy impregnation and double-sided polishing. The mineral composition was identified using polarized light microscopy, and X-ray diffraction (XRD) analysis was performed on 22 bulk samples and <2 μm size fractions using a Neo-Confucianism DMAX-3C diffractometer. The geochemical and information of the diagenetic minerals was evaluated via cathodoluminescence (CL) analysis. The instrument model is CL8200MK5, and the analysis was performed at an accelerating voltage of 30 kV, with a 2Na beam current and a 5-millimeter working distance.

Fifteen double-sided polished thin sections were selected for microthermometric measurements. The homogenization temperatures were measured using a Linkam THMS-600 heating/cooling stage. Only primary fluid inclusions containing both aqueous and hydrocarbon phases were selected from the authigenic minerals to determine their minimum precipitation temperatures.

The chemical composition of the quartz cements was quantitatively determined using electron microprobe analysis (EMPA) with an EPMA-1720JXA-8100 electron microprobe (operating conditions: 15-kV accelerating voltage, 10-nA current, and 1-micrometer beam diameter) at the State Key Laboratory of Oil and Gas Reservoir Geology and Exploitation in Chengdu. The accuracy of experimental analysis is −3–+3%. Four thin sections and 24 sample points were analyzed.

Ion microprobe spots were identified using CL and reflected microscopy, and any areas with well-developed quartz cements (>25 μm wide) were cored from the double-thickness polished thin sections and then packed into one block alongside international NBS-28 standard quartz grains. In situ SIMS oxygen isotopic composition analysis of the detrital quartz and quartz overgrowths was conducted using a CAMECA IMS-1280 ion microprobe with an approximately 15-micrometer diameter beam at the Institute of Geology and Geophysics, Chinese Academy of Sciences (IGGCAS) in Beijing, and the internal precision of $\delta^{18}O$ was ca. 0.2‰ (2 standard deviations, 2SDs) based on 20 measurement cycles [27,28].

4. Results

4.1. Petrographic Description

The sandstone of the Xu2 member is mainly composed of fine- to medium-grained litharenites, sublitharenites, and feldspathic litharenites (Figure 2). The detrital composition is primarily quartz, and the sandstone is relatively poor in feldspar and rich in rock fragments. The rock fragments are predominantly metamorphic and igneous rock fragments, while the feldspars are largely potassic feldspars (orthoclase and microcline) with small quantities of plagioclase. The main components of the interstitial materials are clay matrix and cements such as quartz, chlorite, illite, and carbonates.

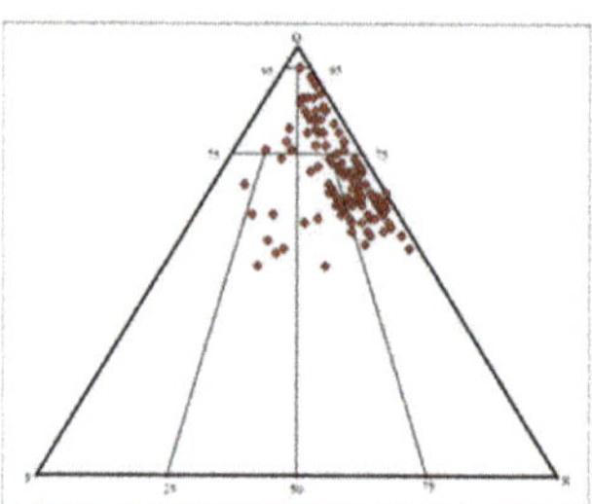

Figure 2. Ternary diagram based on Folk (1968) illustrating the composition of the Xu2 sandstones. Q-detrital quartz grains, F-detrital feldspar grains, R-rock fragments.

4.2. Diagenesis Types and Characteristics

The compaction and pressure dissolution degrees of the deep-buried sandstone in the Xu2 member are high. According to the thin section analysis, line, concave convex, and stylolite contacts occur between the detrital grains, while mica or plastic rock fragment were deformed by compaction (Figure 3a).

K-feldspar dissolution commonly produced secondary pores, which highly improves the reservoir quality of tight sandstones, dissolution was enhanced by fractured grains (Figure 3b,c). Feldspar dissolution mainly occurred along the edges and cleavage surfaces of detrital grains. Some feldspar was largely dissolved to form a honeycomb or hole structure, while the chlorite layer previously covering the detrital grains remains (Figure 3d). It can be seen that carbonate minerals are recrystallized with euhedral outlines and filled in feldspar dissolution pores (Figure 3e,f).

According to the XRD analysis and scanning electron microscopy (SEM) observation, the authigenic clay minerals are primarily chlorite and illite, with no kaolinite. Illite is the most common clay mineral, and it generally occurs as fibrous or hairlike crystals in primary pores as well as within the secondary pores in feldspar grains (Figure 3g). Chlorite is also an important clay mineral in tight sandstone reservoirs, and it usually occurs as rosette-shaped and needle crystals on grain surfaces or in primary pores (Figure 3h).

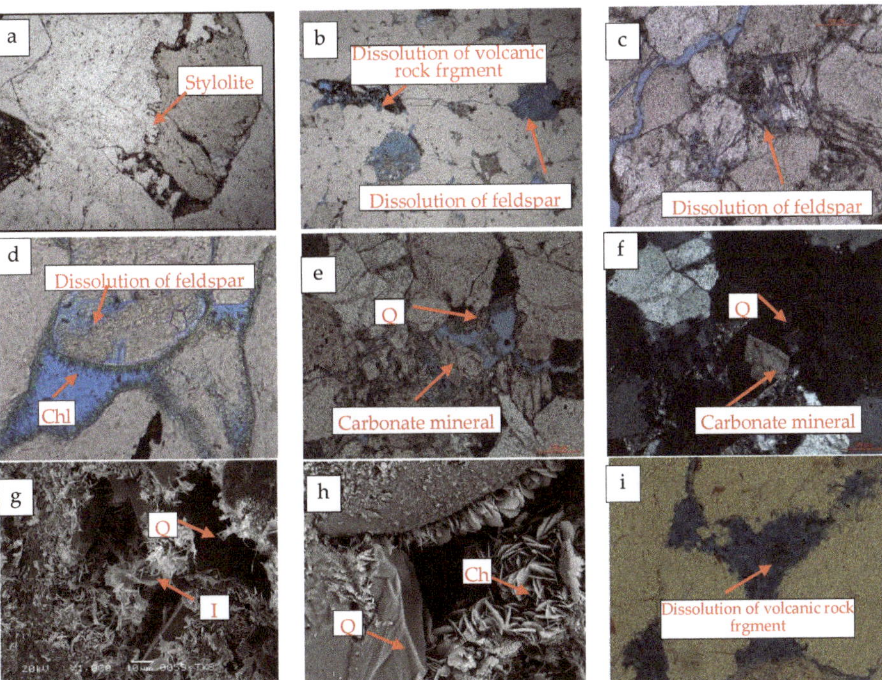

Figure 3. Photomicrographs showing the diagenesis characteristics of the Xu2 sandstones. (**a**) Stylolite contact of the quartz detrital grains, CG561 well, 4926.65 m, plane-polarized light (PPL) analysis; (**b**) Dissolution of feldspar and volcanic rock fragments associated with fractures, intragranular dissolution pores, CX565 well, 4898.31 m, PPL; (**c**) Dissolution of feldspar associated with fractures, intragranular dissolution pores, ZT1 well, 3667.6 m, PPL; (**d**) Dissolution of feldspar and chlorite layer, CG561 well, 4992.9 m, PPL; (**e**) Dissolution of feldspar, carbonate mineral recrystallization, ZT1 well, 3673.05 m, PPL; (**f**) Dissolution of feldspar, carbonate mineral recrystallization, ZT1 well, 3673.05 m, XPL; (**g**) Illite and authigenic quartz cements, X10 well, 4880.24 m, SEM; (**h**) Chlorite-coated and authigenic quartz cements, JH2 well, 3440.42 m, SEM; (**i**) Dissolution of volcanic rock, CG561 well, 4993.54 m.

Quartz cements are volumetrically significant in more mature sandstones and commonly increase with the burial depth. Quartz cements occupy notable amounts of the pore space (7% on average) in the Xu2 sandstones. Three quartz cement formation phases can be distinguished based on the thin section and CL analyses, which are referred to as Q1, Q2, and Q3 (Figure 4). The Q1 quartz cement, which is in direct contact with the quartz grains, is characterized by thin overgrowths surrounding detrital quartz grains and exhibits a slightly gray-black luminescence. It was likely formed earlier than the chlorite- coating (Figure 4a). The subsequently formed Q2 quartz cement is characterized by overgrowths, commonly overlying the Q1 overgrowths, and exhibits a slightly gray-black luminescence, The Q2 quartz cement is volumetrically than Q1 quartz cement (Figure 4b–d). The Q3 quartz cement usually fills intergranular pores and intragranular dissolution pores with euhedral crystals and it exhibits a homogeneous, dark brown luminescence (Figure 4b–d).

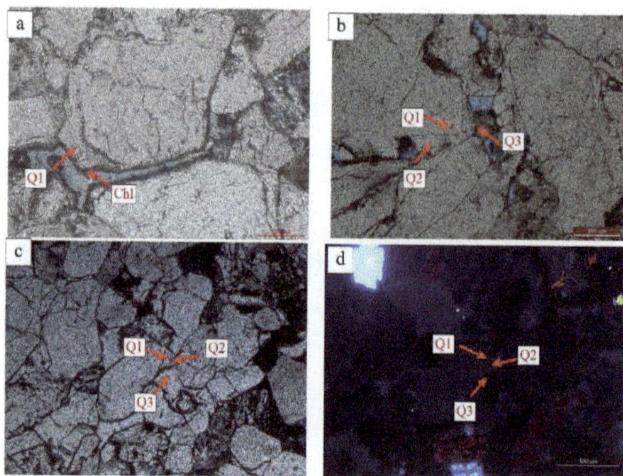

Figure 4. Photomicrographs showing the characteristics of the quartz cements. (**a**) Q1 quartz cement, which precipitated earlier than the chlorite-coated did, JH2 well, 3419 m, PPL; (**b**) Three phases of the quartz cements referred to as Q1, Q2, and Q3, X10 well, 4880.54 m, PPL; (**c**) Three phases of the quartz cements referred to as Q1, Q2, and Q3, JH2 well, 3440.42 m, PPL; (**d**) Three phases of the quartz cements referred to as Q1, Q2, and Q3, JH2 well, 3440.42 m, CL.

4.3. Fluid Inclusion Analysis of the Quartz Cements

Fluid inclusion microthermometric data of the quartz cements were acquired for 39 data points. Primary aqueous fluid inclusions commonly occur along the surface (dust rim) separating the detrital grains from the authigenic overgrowths. The data, however, suggest a trimodal distribution, with three distinct temperature intervals of 64–85, 95–125, and 130–170 °C (Figure 5), corresponding to the Q1, Q2, and Q3 quartz cements, respectively.

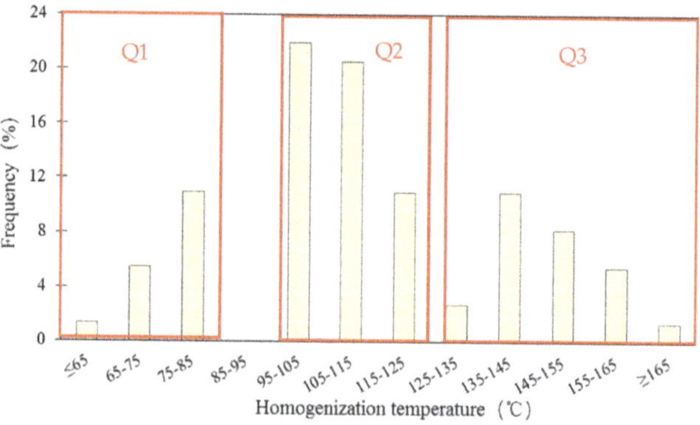

Figure 5. Histograms of the homogenization temperature (Th) in the different phases of the quartz cements in the Xu2 sandstones.

Hydrocarbon fluid inclusions were mainly found in the Q2 and Q3 quartz cements, which emit blue fluorescence under fluorescence microscopy. Similarly, through analysis of the burial history and temperature, source rock maturation and hydrocarbon migration occurred simultaneously with the growth of the Q2 and Q3 quartz cements [23].

4.4. The $\delta^{18}O$ Data of the Quartz Cements and Quartz Grains

The $\delta^{18}O_{\text{(Vienna Standard Mean Ocean Water (VSMOW))}}$ values of the quartz cements in the Xu2 sandstones range from 6.56 to 18.91‰ ($n = 23$) (average 14.33‰) and are higher than those of the detrital quartz grains, which range from 6.56 to 8.21‰ (n = 7). The $\delta^{18}O_{(VSMOW)}$ value of the Q1 quartz cement ranges from 18.3 to 19.05‰ (n = 6), with an average of 18.61‰, and that of the Q2 quartz cement ranges from 15 to 17.99‰ (n = 7), with an average of 16.76‰, while that of the Q3 quartz cement ranges from 12.79 to 15.47‰ (n = 10), with an average of 14.22‰. These data reveal that the $\delta^{18}O_{(VSMOW)}$ value gradually decreases with the distance from the quartz cement to the boundary of the detrital quartz grain (Figure 6, Table 1).

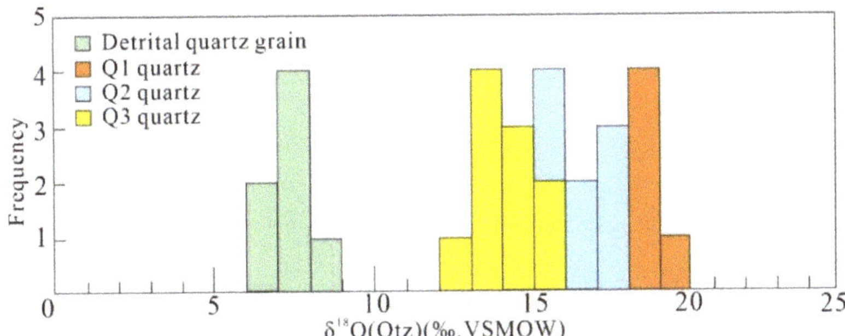

Figure 6. Histograms of the oxygen isotope ratios of the quartz in the Xu2 sandstones measured in situ, including the cements and detrital quartz grains.

Table 1. Oxygen isotope ratios of the quartz in the Xu2 sandstones measured in situ, including the cements and detrital quartz grains.

Point	Sample ID	δ18O‰(SMOW)	2SE	Cements and Detrital Quartz Grains
1	44-2	15.35	0.13	Q3 quartz
2	44-1	17.99	0.18	Q2 quartz
3	44	18.56	0.22	Q1 quartz
4	44-3	8.21	0.18	Detrital quartz grain
5	77-3	12.79	0.12	Q3 quartz
6	77-5	17.97	0.22	Q2 quartz
7	77-2	18.30	0.19	Q1 quartz
8	77-4	7.98	0.30	Detrital quartz grain
9	77-1	7.20	0.20	Detrital quartz grain
10	77	6.76	0.22	Detrital quartz grain
11	x10-1	14.54	0.12	Q3 quartz
12	x10-2	14.84	0.20	Q3 quartz
13	x10-5	15.00	0.18	Q2 quartz
14	x10-4	17.82	0.19	Q2 quartz
15	x10-3	15.99	0.18	Q2 quartz
16	x10-8	18.32	0.20	Q1 quartz
17	x10-9	18.91	0.15	Q1 quartz
18	x10-12	7.77	0.18	Detrital quartz grain
19	x10-10	7.44	0.18	Detrital quartz grain
20	x10-11	6.56	0.19	Detrital quartz grain
21	121-2	14.05	0.08	Q3 quartz
22	121-3	13.98	0.24	Q3 quartz
23	121-4	13.91	0.31	Q3 quartz
24	121	13.78	0.28	Q3 quartz
25	121-7	15.47	0.13	Q3 quartz
26	121-6	13.52	0.18	Q3 quartz
27	121-8	16.44	0.19	Q2 quartz
28	121-9	16.11	0.19	Q2 quartz
29	121-10	19.05	0.23	Q1 quartz
30	121-11	18.45	0.18	Q1 quartz

4.5. Electron Probe Analysis of the Quartz Cements

The trace elements in each quartz cement phase were analyzed with the electron microprobe. A summary of the trace element analysis results is shown in Figure 7, and all the data represent the mass fraction percentage. Notably, aluminum exhibits the largest variation across the quartz cement phases. The Q1 quartz cement contains the lowest Al_2O_3 content with an average of 0.014%. The most volumetrically important Q2 quartz cement exhibits the highest Al_2O_3 content with a mean value of 0.148%. The Q3 quartz is Al-depleted, with a mean value of 0.075%. Moreover, some Q2 and Q3 quartz cements attains a relatively high K_2O content.

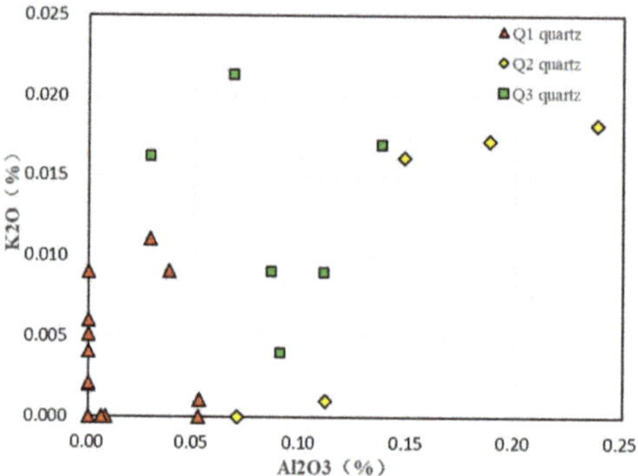

Figure 7. Relationship between Al_2O_3 and K_2O in the different quartz cement phases.

5. Discussion

5.1. Nature and Origin of the Fluid

The $\delta^{18}O_{(VSMOW)}$ values of the quartz cements can be employed to analyze the properties of the diagenetic fluids [9,14,15,27]. The quartz-water fractionation equation of Clayton et al. (1972) [29] is as follows:

$$1000 \ln\alpha_{\text{(quartz-water)}} = 3.38\ (10^6 \times T^{-2}) - 3.40 \quad (1)$$

The Q1 quartz cement precipitated earlier than the chlorite-coated (Figure 4a), with fluid inclusions exhibiting a homogenization temperature ranging from 64 to 85 °C and a calculated paleofluid $\delta^{18}O_{(water)}$ value ranging from −8.16 to −4.05‰ VSMOW, close to the value of meteoric water (−7‰) (Figure 8). The paleofluid may have been influenced by meteoric water. The Q2 quartz cement was mainly formed from 95 to 125 °C, and the calculated paleofluid $\delta^{18}O_{(water)}$ value ranges from −6.63 to −0.09‰ VSMOW (Figure 8). Since the oxygen isotope composition of the formation waters generally evolves to more positive values with increasing burial as a result of water-rock reactions, the recrystallization reactions of the clay minerals and hydrocarbon emplacement tend to result in $\delta^{18}O_{(water)}$ enhancement [4,30–32], and it is inferred that the pore fluid maybe the complex high-temperature mixed fluid remaining after hydrocarbon emplacement and water–rock reactions. The Q3 quartz cement primarily formed from 130 to 170 °C, and the calculated paleofluid $\delta^{18}O_{(water)}$ value ranges from −4.68to 1.54‰ VSMOW (Figure 8). According to the thermal and burial histories (Figure 9), the burial depth of this stage reached 3500 m, and the pore water $\delta^{18}O_{(water)}$ value continuously increased in the middle and late diagenetic stages due to the release of SiO_2 by pressure solution and carbonate mineral recrystallization [4,10,30,31]. It is inferred that the pore fluid in this stage is mainly the mixed fluid remaining after pressure dissolution and water–rock reaction interactions.

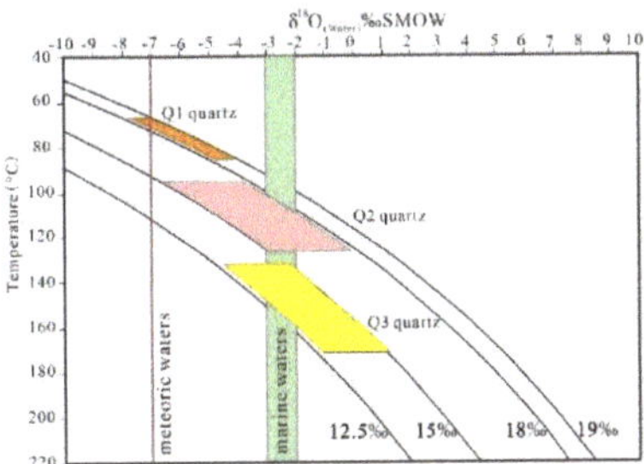

Figure 8. Plot of $\delta^{18}O_{(water)}$ in equilibrium with $\delta^{18}O_{(cement)}$ = 12.5, 15, 18, and 19‰ as a function of the temperature [29], Late Triassic marine waters, $\delta^{18}O_{(water)}$ = approximately −3 to −2‰ [33,34], and meteoric waters, $\delta^{18}O_{(water)}$ = −7‰ [4,35].

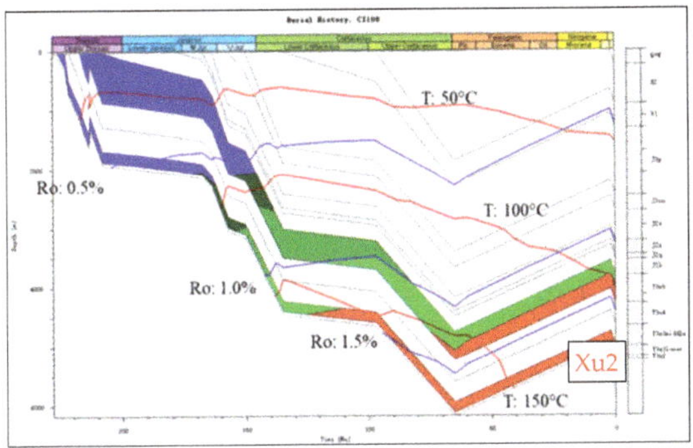

Figure 9. Burial history of the Xu2 member in Western Sichuan Basin.

5.2. Silica Sources

The potential sources of silica for quartz cementation in the sandstones can be summarized as biogenic silica, feldspar dissolution, clay mineral transformation, and pressure dissolution of detrital quartz grains [3]. The Q1 quartz cement was mainly formed at a temperature range of 64 to 85 °C, and the isotopic analysis of the Q1 quartz cement reveals a meteoric water influence. Sandstones containing biogenic silica are generally deposited in marine environments, and they could be potential silica sources in the early diagenetic stage, whereas the Xujiahe Formation is a nonmarine deposit. Therefore, biogenic silica is excluded. At low temperatures, meteoric water dissolves volcanic rock fragments and feldspar, thereby releasing silica [27,36–39] (Figure 3b,i). The fraction of silica released from the dissolution of feldspar minerals can reach as high as 0.43 cm^3 per cm^3 of the dissolved feldspar depending on the original composition and type of the feldspar minerals [40,41]. Therefore, it appears that the dissolution of feldspar and volcanic rock fragments may be the most important source of silica for the Q1 quartz cement.

The Q2 quartz cement was primarily formed from 95 to 125 °C. According to its burial history, the organic acids released from the mature organic matter likely created an acidic environment, which facilitated the precipitation of authigenic quartz [5,42]. Moreover, the acidic fluid dissolved feldspar, and the pore water was enriched with Al^{3+}, K^+, and Na^+; authigenic quartz generally also contains a small amount of trace elements [3]. EPMA analysis reveals that the Q2 quartz cement exhibits a relatively high K_2O content, with the highest Al_2O_3 content (Figure 7). The SEM indicates that the authigenic quartz fills the dissolution pores of the feldspar detrital grains (Figure 10).

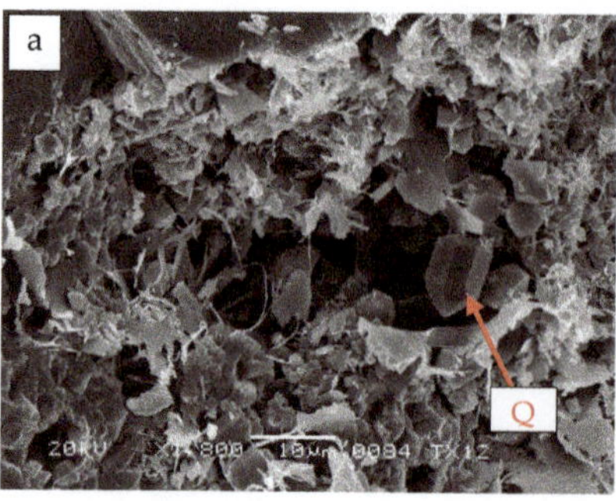

Figure 10. Scanning electron microscopy images showing the characteristics of the quartz cements. (**a**) Dissolution pores of rock fragments filled with authigenic quartz and illite, X10 well, 4880.54 m; (**b**) Dissolution pores of feldspar filled with authigenic quartz, ZT1 well, 3673.05 m.

With an increasing temperature, the smectite in sandstones or adjacent mudstones is gradually transformed into illite or chlorite [3,8,43,44]. The original composition of the smectite may determine whether it is transformed into illite or chlorite. Mg Fe-rich smectite tends to be chloritized, whereas K-, Ca-, and Na-rich smectite tend to be illitized [7,45]. The clay minerals in the Xu2 sandstone are mainly illite and chlorite and are associated with the authigenic quartz (Figure 3g,h). This demonstrates that the transformation of clay minerals could be a potential silica source, but regarding the silica formed in the external shale unit, there must be a suitable transport mechanism for precipitation in the adjacent sandstone reservoir, which is more difficult for the deep-buried thick sandstone of the second member of the Xujiahe Formation. Therefore, the dissolution of feldspar and the transformation of the clay minerals in the Xu2 sandstone appear to have been the most important sources of silica for the Q2 quartz cement.

The Q3 quartz cement formation temperature is generally higher than 130 °C, and the burial depth of the Xu2 sandstone is generally greater than 4500 m. The thin sections reveal that intergranular pressure dissolution and stylolites are common (Figure 3a), indicating that this process could be considered an important source of silica for the Q3 quartz cement. Moreover, mica and illite clay minerals promote quartz grain pressure dissolution to produce free silica [46,47] (Figure 3a). Therefore, pressure dissolution appears to have been the most important source of silica for the Q3 quartz cement.

6. Conclusions

1. Three quartz cement phases are identified based on thin section analysis, CL, and fluid inclusion microthermometry, with three distinct temperature intervals of 56–85, 90–125, and 130–175 °C. Hydrocarbon fluid inclusions are mainly found in the Q2 and Q3 quartz cements, which emit blue fluorescence under fluorescence microscopy.
2. The $\delta^{18}O$ values of the quartz cements range from 6.56 to 18.91‰ (n = 23) (average: 14.33‰) and are higher than those of the detrital quartz grains. The $\delta^{18}O_{(VSMOW)}$ value of the Q1 quartz cement ranges from 18.3 to 19.05‰, that of the Q2 quartz cement ranges from 15 to 17.99‰, while that of the Q3 quartz cement ranges from 12.79 to 15.47‰.
3. The Q1 quartz cement has the lowest Al_2O_3 content with an average of 0.014%. The Q2 quartz cement exhibits the highest Al_2O_3 content with a mean value of 0.148%. The Q3 quartz cement is Al-depleted, with a mean value of 0.075%. Moreover, some Q2 and Q3 quart cements samples achieve a relatively high K_2O content.
4. The dissolution of feldspar and unstable volcanic rock fragments appears to have been the most important source of silica for the Q1 quartz cement. The variations in $\delta^{18}O_{(water)}$ and trace element composition from the Q2 quartz cement to the Q3 quartz cement suggest that hydrocarbon emplacement and water–rock interactions greatly

altered the chemistry of the pore fluid. The dissolution of feldspar, clay mineral transformation, and pressure dissolution appear to have been the most important sources of silica for the Q2 and Q3 quartz cements. The transformation of the clay minerals in the external shale unit was a limited source of silica.

Author Contributions: Conceptualization, Z.L.; formal Analysis, J.W.; data Curation, J.W.; writing—original draft preparation, J.R. and H.W.; writing—review and editing, S.Z. and Z.L. All authors have read and agreed to the published version of the manuscript.

Funding: Our team was supported by the National Science and Technology Major Project [grant number 2016ZX05002-004-010].

Acknowledgments: We thank the anonymous reviewers for their valuable suggestions and kind help in improving the quality of the manuscript, and the Southeast Branch Company of SINOPEC for sample collection and technical support.

Conflicts of Interest: The authors declare no conflict of interest.

References

1. Houseknecht, D.W. Influence of Grain Size and Thermal Maturity on Intergranular Pressure Solution and Quartz Cementation in a Quartz-Rich Sandstone. *J. Sediment. Petrol.* **1984**, *54*, 348–361. [CrossRef]
2. Bjorlykke, K.; Egeberg, P. Quartz Cementation in Sedimentary Basins. *AAPG Bull.* **1993**, *77*, 1538–1548.
3. Worden, R.H.; Morad, S. *Quartz Cementation in Sandstones*; International Association of Sedimentologists, Special Publication, Blackwell Science: Oxford, UK, 2000.
4. Marchand, A.M.E.; Macaulay, C.I.; Haszeldine, R.S.; Fallick, A.E. Pore water evolution in oilfield sandstones: Constraints from oxygen isotope microanalyses of quartz cement. *Chem. Geol.* **2002**, *191*, 285–304. [CrossRef]
5. Mcbride, E.F. Quartz cement in sandstones: A review. *Earth-Sci. Rev.* **1989**, *26*, 69–112. [CrossRef]
6. Lander, R.H.; Larese, R.E.; Bonnell, L.M. Toward more accurate quartz cement models: The importance of euhedral versus noneuhedral growth rates. *Aapg Bull.* **2008**, *92*, 1537–1563. [CrossRef]
7. Bukar, M. *Does Oil Emplacement Stop Diagenesis and Quartz Cementation in Deeply Buried Sandstone Reservoirs*; University of Liverpool: Liverpool, UK, 2013.
8. Van de Kamp, P.C. Smectite-illite-muscovite transformations, quartz dissolution, and Silica release in shales. *Clays Clay Miner.* **2008**, *56*, 66–81. [CrossRef]
9. Harwood, J.; Aplin, A.C.; Fialips, C.; Iliffe, J.E.; Kozdon, R.; Ushikubo, T.; Valley, J.W. Quartz cementation history of sandstones revealed by high-resolution SIMS oxygen isotope analysis. *J. Sediment. Res.* **2013**, *83*, 522–530. [CrossRef]
10. Oye, O.J.; Aplin, A.C.; Jones, S.J.; Gluyasa, J.G.; Bowenb, L.; Orlandc, I.J.; Valley, J.W. Vertical effective stress as a control on quartz cementation in sandstones. *Mar. Pet. Geol.* **2018**, *98*, 640–652. [CrossRef]
11. Walderhaug, O. Kinetic modeling of quartz cementation and porosity loss in deeply buried sandstone reservoirs. *AAPG Bull.* **1996**, *80*, 731–745.
12. Walderhaug, O. Modeling quartz cementation and porosity in Middle Jurassic Brent Group sandstones of the Kvitebjørn field, northern North Sea. *AAPG Bull.* **2000**, *84*, 1325–1339. [CrossRef]
13. Lander, R.H.; Walderhaug, O. Predicting porosity through simulating sandstone compaction and quartz cementation. *AAPG Bull.* **1999**, *83*, 433–449.
14. Hiatt, E.E.; Kurtis, T.K.; Fayek, M.; Polito, P.; Holk, G.J.; Riciputi, L.R. Early quartz cements and evolution of paleohydraulic properties of basal sandstones in three Paleoproterozoic continental basins: Evidence from in situ $\delta^{18}O$ analysis of quartz cements. *Chem. Geol.* **2007**, *238*, 19–37. [CrossRef]
15. Hyodo, A.; Kozdon, R.; Pollington, A.D.; Valley, J.W. Evolution of quartz cementation and burial history of the Eau Claire Formation based on in situ oxygen isotope analysis of quartz overgrowths. *Chem. Geol.* **2014**, *384*, 168–180. [CrossRef]
16. Longstaffe, F.J.; Ayalon, A. *Oxygen-Isotope Studies of Clastic Diagenesis in the Lower Cretaceous Viking Formation, Alberta: Implications for the Role of Meteoric Water*; Geological Society, Special Publications: London, UK, 1987; Volume 36, pp. 277–296.
17. Guo, Z.W.; Deng, K.L.; Han, Y.H. *Formation and Evolution of the Sichuan Basin*; Geological Publishing House: Beijing, China, 1996; pp. 1–200, (In Chinese with English Abstract).
18. Gao, B.; Tian, F.; Pan, R.; Zheng, W.H.; Li, R.; Huang, T.J.; Liu, Y.S. Hydrothermal Dolomite Paleokarst Reservoir Development in Wolonghe Gasfield, Sichuan Basin, Revealed by Seismic Characterization. *Water* **2020**, *12*, 579. [CrossRef]
19. Yue, D.L.; Wu, S.H.; Xu, Z.Y.; Xiong, L.; Chen, D.X.; Ji, Y.L. Reservoir quality, natural fractures, and gas productivity of upper Triassic Xujiahe tight gas sandstones in western Sichuan Basin, China. *Mar. Pet. Geol.* **2018**, *89*, 370–386. [CrossRef]
20. Ni, Y.Y.; Dai, J.X.; Tao, S.Z.; Wu, X.Q.; Liao, F.R.; Wu, W.; Zhang, D.J. Helium signatures of gases from the Sichuan Basin, China. *Org. Geochem.* **2014**, *74*, 33–43. [CrossRef]

21. Lai, J.; Wang, G.W.; Ran, Y.; Zhou, Z.L. Predictive distribution of high-quality reservoirs of tight gas sandstones by linking diagenesis to depositional facies: Evidence from Xu-2 sandstones in the Penglai area of the central Sichuan basin, China. *J. Nat. Gas Sci. Eng.* **2015**, *23*, 97–111. [CrossRef]
22. Tao, S.Z.; Zou, C.N.; Mi, J.K.; Gao, X.H.; Yang, C.; Zhang, X.X.; Fan, J.W. Geochemical comparison between gas in fluid inclusions and gas produced from the Upper Triassic Xujiahe Formation, Sichuan Basin, SW China. *Org. Geochem.* **2014**, *74*, 59–65. [CrossRef]
23. Luo, L.; Meng, W.B.; Gluyas, J.; Tan, X.F.; Gao, X.Z.; Feng, M.S.; Kong, X.Y.; Shao, H.B. Diagenetic characteristics, evolution, controlling factors of diagenetic system and their impacts on reservoir quality in tight deltaic sandstones: Typical example from the Xujiahe Formation in Western Sichuan Foreland Basin, SW China. *Mar. Pet. Geol.* **2019**, *103*, 231–254. [CrossRef]
24. Zhang, S.L.; Lv, Z.X.; Wen, Y.; Liu, S.B. Origins and Geochemistry of Dolomites and Their Dissolution in the Middle Triassic Leikoupo Formation, Western Sichuan Basin, China. *Minerals* **2018**, *8*, 289. [CrossRef]
25. Zhang, L.; Guo, X.S.; Hao, F.; Zou, H.Y.; Li, P. Lithologic characteristics and diagenesis of the upper triassic Xujiahe Formation, yuanba area, northeastern Sichuan Basin. *J. Nat. Gas Sci. Eng.* **2016**, *35*, 1320–1335. [CrossRef]
26. Xu, C.M.; Gehenn, J.M.; Zhao, D.H.; Xie, G.Y.; Teng, M.K. The fluvial and lacustrine sedimentary systems and stratigraphic correlation in the Upper Triassic Xujiahe Formation in Sichuan Basin, China. *AAPG Bull.* **2015**, *99*, 2023–2041. [CrossRef]
27. Yuan, G.H.; Cao, Y.C.; Gluyas, J.; Cao, X.; Zhang, W.B. Petrography, fluid-inclusion, isotope, and trace-element constraints on the origin of quartz cementation and feldspar dissolution and the associated fluid evolution in arkosic sandstones. *AAPG Bull.* **2018**, *102*, 761–792. [CrossRef]
28. Li, X.H.; Li, Z.X.; He, B.; Li, W.X.; Li, Q.L.; Gao, Y.Y.; Wang, X.C. The Early Permian active continental margin and crustal growth of the Cathaysia Block: In situ U-Pb, Lu-Hf and O isotope analyses of detrital zircons. *Chem. Geol.* **2012**, *328*, 195–207. [CrossRef]
29. Clayton, R.N.; O'Neil, J.R.; Mayeda, T.K. Oxygen isotope exchange between quartz and water. *J. Geophys. Res.* **1972**, *77*, 3057–3067. [CrossRef]
30. Aplin, A.C.; Warren, E.A. Oxygen isotopic indications of the mechanisms of silica transport and quartz cementation in deeply buried sandstones. *Geology* **1994**, *22*, 847–850. [CrossRef]
31. Wilkinson, M.; Crowley, S.F.; Marshall, J.D. Model for the evolution of oxygen isotope ratios in the pore fluids of mudrocks during burial. *Mar. Pet. Geol.* **1992**, *9*, 98–105. [CrossRef]
32. Girard, J.P.; Munz, I.A.; Johansen, H.; Hill, S.; Canham, A. Conditions and timing of quartz cementation in Brent reservoirs, Hild Field, North Sea: Constraints from fluid inclusions and SIMS oxygen isotope microanalysis. *Chem. Geol.* **2001**, *176*, 73–92. [CrossRef]
33. Muttoni, G.; Mazza, M.; Mosher, D.; Katz, M.E.; Kent, D.V.; Balini, M. A Middle-Late Triassic (Ladinian-Rhaetian) carbon and oxygen isotope record from the Tethyan Ocean. *Palaeogeogr. Palaeoclimatol. Palaeoecol.* **2014**, *399*, 246–259. [CrossRef]
34. Korte, C.; Kozur, H.W.; Veizer, J. $\delta^{13}C$ and $\delta^{18}O$ values of Triassic brachiopods and carbonate rocks as proxies for coeval seawater and palaeotemperature. *Palaeogeogr. Palaeoclimatol. Palaeoecol.* **2005**, *226*, 287–306. [CrossRef]
35. Liu, S.B.; Huang, S.J.; Shen, Z.M.; Lv, Z.X.; Song, R.C. Diagenetic fluid evolution and water-rock interaction model of carbonate cements in sandstone: An example from the reservoir sandstone of the Fourth Member of the Xujiahe Formation of the Xiaoquan-Fenggu area, Sichuan Province, China. *Sci. China Earth Sci.* **2014**, *57*, 1077–1092. [CrossRef] (In Chinese with English Abstract)
36. Hurst, A.; Irwin, H. Geological modelling of clay diagenesis in sandstones. *Clay Miner.* **1982**, *17*, 5–22. [CrossRef]
37. Franca, A.B.; Aranujo, L.M.; Maynard, J.B.; Potter, P.E. Secondary porosity formed by deep meteoric leaching: Botucatu eolianite, southern South America. *AAPG Bull.* **2003**, *87*, 1073–1082. [CrossRef]
38. Liu, Y.F.; Hu, W.X.; Cao, J.; Wang, X.L.; Tang, Q.S.; Wu, H.G.; Kang, X. Diagenetic constraints on the heterogeneity of tight sandstone reservoirs: A case study on the Upper Triassic Xujiahe Formation in the Sichuan Basin, southwest China. *Mar. Pet. Geol.* **2018**, *92*, 650–669. [CrossRef]
39. Sun, H.T.; Zhong, D.K.; Zhan, W.J. Reservoir characteristics in the Cretaceous volcanic rocks of Songliao Basin, China: A case of dynamics and evolution of the volcano-porosity and diagenesis. *Energy Explor. Exploit.* **2019**, *37*, 607–625. [CrossRef]
40. Leder, F.; Park, W. Porosity reduction in sandstone by quartz overgrowth. *Aapg Bull.* **1986**, *70*, 1713–1728.
41. Morad, S.; AlDahan, A.A. A SEM study of diagenetic kaolinitization and illitization of detrital feldspar in sandstones. *Clay Miner.* **1987**, *22*, 237–243. [CrossRef]
42. Zaid, S.M. Provenance, diagenesis, tectonic setting and geochemistry of Rudies sandstone (Lower Miocene), Warda Field, Gulf of Suez, Egypt. *J. Afr. Earth Sci.* **2012**, *66–67*, 56–71. [CrossRef]
43. Perry, E.A.; Hower, J. Burial diagenesis in Gulf Coast pelitic sediments. *Clay Clay Miner.* **1970**, *18*, 167–177. [CrossRef]
44. Lynch, F.L.; Mack, L.E.; Land, L.S. Burial diagenesis of illite/smectite in shales and the origins of authigenic quartz and secondary porosity in sandstones. *Geochim. Cosmochim. Acta* **1997**, *61*, 1995–2006. [CrossRef]
45. Chang, H.K.; Mackenzie, F.T.; Schoonmaker, J. Comparisons between the diagenesis of dioctahedral and trioctahedral smectite, Brazilian offshore basins. *Clays Clay Miner.* **1986**, *34*, 407–423. [CrossRef]
46. Oelkers, E.H.; Bjorkum, P.A.; Murphy, W.M. A petrographic and computational investigation of quartz cementation and porosity reduction in North Sea sandstones. *Am. J. Sci.* **1996**, *296*, 420–452. [CrossRef]
47. Walderhaug, O. Kaolin-Coating of Stylolites, Effect on Quartz Cementation and General Implications for Dissolution at Mineral Interfaces. *J. Sediment. Res.* **2006**, *76*, 234–243. [CrossRef]

MDPI
St. Alban-Anlage 66
4052 Basel
Switzerland
Tel. +41 61 683 77 34
Fax +41 61 302 89 18
www.mdpi.com

Water Editorial Office
E-mail: water@mdpi.com
www.mdpi.com/journal/water

www.ingramcontent.com/pod-product-compliance
Lightning Source LLC
LaVergne TN
LVHW070642170726
843515LV00004B/308
* 9 7 8 3 0 3 6 5 5 5 4 5 4 *